Responsible Mining

Mining has negative environmental and social impacts, but it can also be responsible. Corporations have little impetus to act responsibly, however, without being held to account by an informed and active public, and by strong institutions and governments which not only create but also enforce legislation. Yet what does such practice look like?

This book shows how the concept of responsible mining is based on five key principles: holistic assessment, ethical relationships, community-based agreements, appropriate boundaries and good governance. Together, these pillars circumscribe global best practice and innovative ideas to catalyse new and improved approaches to a sustainable mining industry. The author argues that these practices are critical to the future viability and social acceptability of the global mining industry and draws on a range of case studies from Australia, Canada, Central Asia, Papua New Guinea and west Africa.

The role of informed communities, governments and civil societies in holding the industry to account to achieve responsible mining is assessed. The book explains how companies judge what effects they may have on communities and investigates ways to improve the prediction and prevention of such impacts and to provide clearer, more meaningful public communication. It offers alternatives to common 'corporate social responsibility' practices in which mining companies adopt roles which are usually the remit of government. Ultimately, it looks to the future, exploring the essential pathways towards responsible mining in the twenty-first century.

Sara Bice is Director, Research Translation, Melbourne School of Government, the University of Melbourne, Australia.

Routledge Studies of the Extractive Industries and Sustainable Development

African Artisanal Mining from the Inside Out
Access, norms and power in Congo's gold sector
Sarah Geenan

Mountain Movers
Mining, sustainability and the agents of change
Daniel M. Franks

Responsible Mining
Key principles for industry integrity
Sara Bice

"This is a timely and perceptive book. It comes at a time of anxious self-reflection in the extractive sector, with many bloated assets revealed stranded in the mudflats of a declining financial tide. Social ledgers are equally discriminating. In this context, Sara Bice has produced a contemporary and erudite analysis of the social dimensions of mining. Her work rejects debate based on ideology and polemic, it simply asks 'is it possible for miners to be socially responsible and acceptable?'. The answer is 'yes', and Sara proceeds to explain why, revealing the new professional discipline of 'social performance'. This is a discipline that turns its back on out-dated stratagems such as philanthropy, advocacy and brokerage, and instead adopts a cold-nosed methodological approach. As many mining businesses flounder in the glare of social scrutiny, others quietly succeed. This book helps you understand why."

Bruce Harvey, Director, Resolution 88 and University of Queensland, Australia

Responsible Mining

Key principles for industry integrity

Sara Bice

Routledge
Taylor & Francis Group
LONDON AND NEW YORK

earthscan
from Routledge

First published 2016
by Routledge
2 Park Square, Milton Park, Abingdon, Oxon OX14 4RN

and by Routledge
711 Third Avenue, New York, NY 10017

Routledge is an imprint of the Taylor & Francis Group, an Informa business

British Library Cataloguing-in-Publication Data
A catalogue record for this book is available from the British Library

Library of Congress Cataloging in Publication Data
Names: Bice, Sara.
Title: Responsible mining : key principles for industry integrity / Sara Bice.
Description: London ; New York : Routledge, 2016. | Series: Routledge studies of the extractive industries and sustainable development | Includes bibliographical references and index.
Identifiers: LCCN 2015050668| ISBN 9781138788244 (hbk) | ISBN 9781138788275 (pbk) | ISBN 9781315765679 (ebk)
Subjects: LCSH: Mineral industries--Moral and ethical aspects. | Mineral industries--Social aspects. | Mineral industries--Environmental aspects. | Social responsibility of business. | Sustainable development.
Classification: LCC HD9506.A2 B4175 2016 | DDC 622.068/4--dc23
LC record available at http://lccn.loc.gov/2015050668

ISBN: 978-1-138-78824-4 (hbk)
ISBN: 978-1-138-78827-5 (pbk)
ISBN: 978-1-315-76567-9 (ebk)

Typeset in Goudy
by Saxon Graphics Ltd, Derby

For Pop

Contents

Acronyms and abbreviations

ASX	Australian Stock Exchange
BSR	Business for Social Responsibility
CBA	community-based agreement-making
CEO	chief executive officer
CFO	chief financial officer
CFP	corporate financial performance
CIBA	community impact and benefit agreement
CSP	corporate social performance
CSR	corporate social responsibility
EEO	equal employment opportunity
EIA	environmental impact assessment
EIR	Extractive Industries Review
EITI	Extractive Industries Transparency Initiative
ERM	Environmental Resources Management
ESG	environmental, social and governance
ESIA	environmental and social impact assessment
FIFO	fly-in/fly-out
GMI	Global Mining Initiative
GRI	Global Reporting Initiative
GRI G3	Global Reporting Initiative sustainability reporting guidelines, version 3
GRI G4	Global Reporting Initiative sustainability reporting guidelines, version 4
HIA	human impact assessment
HRIA	human rights impact assessment
IAIA	International Association for Impact Assessment
IBA	Impact and benefit agreement
ICMM	International Council on Mining and Metals
IFC	International Finance Corporation
IIED	International Institute for Environment and Development
IRMA	Initiative for Responsible Mining Assurance
ISO	International Organization for Standardization
KPIs	key performance indicators
MCA	Minerals Council of Australia

MMSD	Mining, Minerals and Sustainable Development
MoU	Memorandum/a of Understanding
NGO	non-government organisation
NRM	Natural Resources Management
OECD	Organisation for Economic Cooperation and Development
OHS	Occupational health and safety
SIA	social impact assessment
SRI	socially responsible investment
UN	United Nations
UNGC	United Nations Global Compact
WBCSD	World Business Council for Sustainable Development

Acknowledgements

I love reading acknowledgements. So much so that, even when they appear at the front of a book, I'll often save them until last. Perhaps it is a legacy of growing up in the American South, my grandmother's insistence on 'please' and 'thank you' as vital to life as breathing. I like having engaged with a text and then returning – with some greater understanding of the author and her endeavours – to discover those behind-the-scenes barrackers whose work, intelligence, insights, advice, experiences, love, meal delivery, spreadsheet wrangling, open conversations, childminding, financial backing, bottomless encouragement, companionable drinking, proofreading, critiques and belief in the project made the text possible.

Thank you to the Melbourne School of Government and The University of Melbourne for providing the space and time to enable completion of the book. To my MSoG colleagues and our Director, Professor Helen Sullivan, thank you for creating such a supportive and collegial research environment. Thank you to the University of Melbourne Writers' Centre for the manuscript sponsorship and to the Faculty of Arts for an Early Career Researcher Grant to support completion of this book. Shout out to my Writers' Centre book group: Brig, Annie and Suzette – it's here! To my research assistants: Dr Margherita Matera for endless patience coding sustainability reports. Chapter 3 is indebted to her efforts and those of Ms Colette Einfeld who made sense of ten years of data. To Ms Joanna Hanley whose 11th hour efforts are the only reason this book is here now. Finally. Phew.

To my former colleagues at the Australian Centre for Corporate Social Responsibility and Director Leeora Black, thank you for early support of my career in CSR and the opportunities that afforded. My time at Oxfam Australia was equally formative and committed my attention to the global mining and extractives industry. I am grateful to have had that experience at such an early stage.

This book would not exist without the generosity of so many people in mine-affected communities and within the industry who gave their time without incentive and, in many cases, bravely and openly shared their stories. Thank you and I hope this book makes some difference to all our futures.

To my colleagues at the International Association for Impact Assessment, especially my co-chair Corporate Stewardship and Risk Management, Rita Sully, your dedication to the practice is exceptional and I hope this text helps to progress assessments further.

Thank you to the numerous and generous researchers who have inspired me and always offered a listening ear and enthusiasm for my projects. You are too many to name properly, but to make a start: Dr Robert Boutilier, Dr Ana Maria Esteves, Dr Daniel Franks, Professor Fiona Haines, Associate Professor Deanna Kemp, Professor Tim Marjoribanks, Dr Kieren Moffat, Dr Angus Morrison-Saunders, Dr Jenny Pope, Professor Mike Sandiford and Professor Frank Vanclay. Thank you also to the Centre for Social Responsibility in Mining at the Sustainable Minerals Institute, The University of Queensland, for involving and hosting me regularly over many years, and to the University of Chicago for an invaluable doctoral fellowship that informed this book.

My parents, Sue and Alan Bice, are two of the nicest people you could meet and they have only ever said 'yes' to opportunities for me. Thank you. To my sisters, Ali and Lauren, for continuing to play Boggle. Wendy Coates, indefatigable reader, dear friend and grandmother extraordinaire – thank you. To my husband, Hamish Coates, and to Annabel and Momo, my life is incredibly sunny with you in it. And Annabel, without you this book would be a year younger and it's totally worth it.

As they say, all faults remain my own.

Prologue

Machetes wielded by lean, brown arms thwhacked the long grass in front of me. The now open path twisted to the left and as our small hiking party mounted the rise, we saw the river we had come to see. Tens of metres across, concrete grey water lathered over smooth rocks. Just downstream, the water churned down a cataract, sweeping loose branches angrily off into the jungle below. In calmer pools near the river bank, light refracted prisms off an oily residue. A few plastic bits and bobs washed up here and there. Augustine and Morris began sawing at a sturdy sapling near the bank, gently pushing until the tree yielded and fell. Two other men came to help lift and manoeuvre the trunk across several rocks, the tree top now resting on a boulder in the centre of the river. 'A bridge!' Augustine said. 'Riiiiggghhtt', I thought glancing furtively around for some alternative means aligned with Western safety standards – a nice punt, perhaps – which would carry me safely to the other side. But they had already started across. I took a deep breath and stepped out over the chute.

It was 2005 and just your average, humid, sticky day in the Papua New Guinean jungle, north of Port Moresby and west of the Owen Stanley Ranges. On assignment for an international development agency, my colleague and I were trekking along the Tolukuma River, downstream of the Tolukuma Gold Mine. Since 1994, and with the approval of the Papua New Guinean Government, the mine had been dumping its tailings – the waste rock left over when you extract the valuable minerals from an ore – into the river. Convenient. Cost-effective. Environmentally and socially catastrophic.

We travelled to the remote site to collect blood, hair and water samples to determine whether people were being exposed to harmful amounts of heavy metals or chemicals as a result of riverine mine tailings disposal. My colleague was the real scientist. I was just the social scientist. But I proved adept at barbering and placing clipped hair in a plastic bag. And I can fill a vial with water, so, for purposes of this trip I was a bona fide wannabe scientist. My real job was to interview as many villagers from the local area as possible to understand better the extent to which the mine's operations were affecting their traditional, subsistence lifestyles, their health, their families and their hopes for the future.

Earlier that morning, we breakfasted in Fané village, a place unknown to me just a few days prior. Yet Fané's strangeness from my usual existence made it quickly, intensely familiar. I internalized every detail of a place I knew I may never see again, that most will never see. My first sighting of life at Fané was of a lone mother bathing her toddler under a trickling mountain cascade, clear water pouring off rocks worn smooth. The village itself consisted of six thatched huts, some larger than others, set widely apart on a large patch of cleared dirt. A community garden rested along a verdant slope behind the huts. Pit toilets to your right. A few dogs lolled in the dirt and groups of children chased each other in looping figure eights. Fané was a village stopped in time.

Each morning, in the pre-dawn light, before the thick heat descended, women and men made their way to the river. The women crossed, as I had, on makeshift bridges or wading, thigh deep, across well-known shallows to reach their garden beds, laden billums hanging from their foreheads across their backs. Later, they would complain to me of rose-red sores on their legs, certain they bloomed from daily exposure to the polluted river water. The men released fishing nets with little hope of a catch, as they had watched fish stocks deplete over the years. The river was no longer used for drinking water. Villagers didn't know for sure what the changes they had witnessed to the lifeblood of their community meant, but they knew the river was no longer theirs to tend, to fish, to consume. In all its changes, however, it remained theirs. So, Fané persisted. But Fané was under threat.

In the year before, I worked with Australian indigenous women whose daily lives and cultural practices were affected by mining operations in the Pilbara region, western Australia. Although thousands of kilometres away, in a vastly different cultural setting, strong similarities between the experiences of these communities emerged. I would later see similar mine-related themes highlighted at villages in western Fiji. Each mine-side community told of how daily life, wellbeing, gender roles, cultural traditions, the environment and local economy were affected by the mine. I was sometimes astounded at the degree of impact on these communities and frequently disheartened at how little is known about their experiences.

In recent years, I have worked more directly with mining companies which recognize that they must mitigate and address their social and environmental impacts more proactively. While many people remain cynical about the sincerity of these companies' actions – and I myself hold certain reservations – the operation site employees with whom I have come into contact demonstrate earnest concern about the communities in which they are working. While this is not to say that every miner, plant supervisor or engineer is concerned about the social implications of her/his work, there appears to be a critical mass of individuals emerging within the mining community who believe their work does not have to be exploitative or neglectful. Improved monitoring and technologies can reduce mining's potentially enormous environmental impacts. Employee education and training can improve workers' health and forestall injuries and fatalities. Awareness of mining's extensive socio-cultural implications means more and more companies are working with local communities to attempt to address these issues even before they begin. Activities of recent years demonstrate that while mining can never be sustainable (in the truest sense of the word), it holds the potential to be responsible. Yet much work remains to be done. The ideas presented in this book are strongly motivated by my experiences with mine-affected communities and suggest a way forward, towards responsible mining.

1 Introduction

Each year more than 4.2 billion tonnes of minerals are extracted from the earth (International Council on Mining and Metals 2012). This is enough material to construct 4,200 Eiffel Towers,[1] covering an area greater than China's Forbidden City.[2] From the Amazon Basin to the Australian Outback, from the Rocky Mountains to the Democratic Republic of Congo we are ceaselessly and intensively depleting our finite resources. Meanwhile, global goods consumption is on the rise, fuelled by a rapidly expanding middle class, many of whom are exiting poverty for the first time in history. In India, for example, over 291 million people will transition from economic destitution into a consumer class by 2025 (Ablett et al. 2007). Similar shifts in China mean that up to 330 million cars could jam the Asian giant's streets by 2020 (Fangfang 2013). Despite economic downturns in developed, Western countries, personal consumption continues apace. In 2012 Americans spent US$1.25 trillion on household equipment, recreational vehicles, cars and car parts alone. These consumer products require raw materials – many of them mined – and energy to fuel production and further consumption. Demand for coal is up globally in all nations except the United States where it is being swapped out for cheaper shale gas (International Energy Agency 2012). China's demand for iron ore propelled record trade levels in 2011 (OECD 2012). Our appetite for modern life and the resources which sustain it is rapacious.

Within the next 200 years, the minerals on which we now rely will be mostly dug up and used (Moyer and Storrs 2010); burned for fuel, lying in long-forgotten rubbish heaps, shining in an urban terrain of high-rises. But the landscape is not all bleak. If we accept that mountains must be disappeared into basins, that ores must be extracted, dirt turned to gold or girders, we can also expect that it is done in a way which is more palatable. Mining can be responsible. It seems like the greatest irony. And it is all too easy to say, 'But it's bad. We must stop it. We must oppose it.' To do so is to deny the modern lives we lead. This book, therefore, does not call for an end to the mining story, but a new and better narrative. A pragmatic but conscious and ethical approach.

Every material aspect of our daily lives is facilitated by mining. Yet many of us never realize, or prefer to leave unacknowledged, the deep connections we hold to an industry which regularly destroys pristine environments, consumes resources which can never be replenished and shapes the prospects of future generations. To reject mining is to reject modern life. To accept it is to acquiesce to certain environmental, social, and economic compromises.

Mining can be responsible. But corporations have little impetus to act responsibly without being held to account by an informed and active public, by strong institutions and governments which not only create but enforce legislation. To do this, corporations, the public, our institutions and governments must understand the possibilities and perimeters of responsible mining. Responsible mining must be comprehensive in its approach and in the concerns it encompasses. From exploration to closure, from environment to community, from corporate governance to financial transparency, this book articulates five pillars to undergird a viable, ethical and accountable mining industry. Taken together, the pillars of holistic assessment, ethical decision-making, community-based agreement-making, appropriate boundaries and good governance hold the potential to transform the mining industry and its impacts. So, what does responsible mining look like? And what should we expect?

The lay of the land

Mining is an exercise of astonishing scale. Your average mine pit is constructed in a series of 'benches', each about the height of a five storey building, stacked in threes or fours with safety benches and haulage roads running along crests in between. Carved from rich, brown earth these benches appear supple and circumfluent, looping down to the grey pit floor. Their gentle grade and terraced edges belie the careful engineering which makes them possible. Each slope is painstakingly planned and measured by engineers speaking a tribal language of haulage types, ore body varieties, overburden amounts and local geology. The world's largest mine pit at Bingham Canyon, Utah is 4.4 km wide and 1.2 km deep. If it were a football stadium, it could seat nine million cheering fans. The roar would be deafening. From above, a mine pit appears as a still, contemplative organism, a great grey-brown amoeba alive with a cellular thrum of human activity. It is so wide, your eyes must first work to take it all in, to grasp the expanse, before waking up to specks of movement. The pit is dotted with busy, yellow machinery. Excavators with buckets the size of a small house shift tonnes of dirt, scoop after colossal scoop. Dump trucks carry hundreds of tonnes of overburden along the haulage roads, up and out. It is only when one of these trucks rumbles nearby that you begin to grasp the size and depth of the pit. The truck, which in the pit appears like a childhood Tonka toy, rolls on tyres two lanes wide and more than twice as tall as your average miner, each costing more than a Mercedes Roadster. The drivers of these gargantuan gas guzzlers are specially trained and usually earn six figures a year. Big pits mean big business.

The global mining industry generated record profits of US$133 billion in 2011. The top 40 global miners paid out dividends in excess of US$32 billion (PricewaterhouseCoopers 2012), more than the purchasing power parity GDP of many small-to-medium sized countries (US Central Intelligence Agency 2013).[3] Mining workers in Australia, a country home to one of the world's largest and longest contemporary mining booms, earn average weekly salaries over US$831 (AU$1,000) more than other workers. In the Land Down Under, mining is a cornerstone of the economy, representing the number one export industry, contributing an estimated seven percent of GDP and comprising almost 2,500 firms paying out US$14.97 billion (AU$18 billion) in wages to 180,000 individuals (Australian Bureau of Statistics 2010; Minerals Council of Australia 2012).

Yet Australia is unusual in its position as an economically developed country whose mining industry remains a primary economic pillar. Although economists argue the Aussie boom is slowing, production and export will continue for years to come (OECD 2013). Australia's assuredness in the viability of its mining industry is based both on rising global demands and on confidence in measures of the remaining quantity and quality of resources, especially iron ore (Geoscience Australia 2013). Other developed nations face depleted resources and dwindling ore quality. Between 1900 and 1960, American mines accounted for between 30 to 40 percent of total global mining (International Council on Mining and Metals 2012). Today, mining in America accounts for less than 10 percent of the global mining industry, extracting mainly coal. The US' neighbour to the north was also once a dominant miner, offering a wide variety of resources. Despite a declining contribution to total global mining output, Canadian mining is persistent and contributed about eight percent of the country's GDP in 2012 (Statistics Canada 2013). Yet explorers remain hopeful, spending an estimated US$134.14 million (C$154 million) searching for minerals in 2010. The investment may pay off, as scientists believe they have identified one of the world's largest, untapped gold-copper resources in British Columbia (Kosich 2010).

The slow-down in developed countries' mining industries is partly due to declining quality of remaining resources. In the United States today, it takes several hundred metric tonnes of ore to produce one metric tonne of copper, resulting in mountains of waste rock up to 400 hectares across (United States Environmental Protection Agency 2012). Since the Great Depression, the amount of waste necessary to produce US copper has increased because the high quality ores – those bodies of rock containing the greatest mineral concentrations – have been tapped, leaving mostly 'low grade ores' which hold lesser amounts of minerals, harder to extract (Mudd 2009). This means more rocks must be broken, more chemicals used, more ore processed and more waste produced to realize the same amount of materials. The stories are similar for minerals including nickel, gold and zinc (International Council on Mining and Metals 2012). And these are only the resources most people recognize by name.

Rare earth minerals in which there was previously limited interest, such as europium, erbium, gadolinium and terbium, are today in high demand. While the names may seem otherworldly, these are the very elements with which we are most likely to have daily contact. They allow our iphone screens to function, provide the finish for eyeglasses and mirrors, facilitate the miniaturization necessary for our computers, and make our light bulbs more energy efficient (Haxel et al. 2002). These minerals are in ubiquitous use across every continent, yet an estimated 95 percent of global supply comes from China (Tse 2011). Following the Global

Financial Crisis, China implemented export limits on its rare earth supply, restricting worldwide access. Not only did China's move cause prices to skyrocket, it forced Western governments to question their reliance on a foreign provider as the source of materials essential to maintaining their military and technological dominance (Brennan 2013). Mining is not just geological. It is geopolitical.

Mining is also no longer bound to the earth. Miners know terrestrial mineral resources are limited, and their attention is turning to new technologies and new frontiers. Our ocean floor is littered with untapped resources. In 2011 the Papua New Guinean government granted Canadian mining company Nautilus Minerals Inc. the world's first deep sea mining lease (Roche 2014). With the approval of the International Seabed Authority, Nautilus plans to use underwater digging and vacuum technologies to suck mineral rich 'polymetallic nodules' to the surface. Although knowledge of these minerals' locations is not new, it is only now that the technology exists to remove them cost-effectively. Defence behemoth Lockheed Martin, for example, will adapt its aerospace and underwater technologies to extract resources from approximately four kilometres beneath the ocean's surface.[4] Proponents argue the deep sea will provide the answer to environmental and social concerns about mining by taking the process away from the communities and landscapes it effects (Roche 2014). Yet the questions raised by mining an area which is inaccessible to most and mysterious to many are substantial. Some answer that space will provide the solution, with metal-laden asteroids floated home to earth, perhaps in a parachuting extravaganza à la Felix Baumgartner (Sonter 1997). Extraterrestrial mining smacks of science fiction. Yet NASA scientists debate its costs, not its possibility (Cohen 2013). The prospects seem endless and are equal only to our imaginations and to our neverending thirst for convenient, high tech, comfortable, beautiful, modern life.

Mineral wealth and its discontents

Back on land, the contemporary 'Eureka!' is being shouted most often in remote regions previously unreachable to mining technology. Today, developing countries generate more than one-fifth of total global mineral production (International Council on Mining and Metals 2012). Europe (excluding Russia) and the United States generate only 3.5 percent and 4.2 percent, respectively, by comparison. The shift of mining activity from the developed to the developing world makes mineral extraction an increasingly important component of developing economies. In countries like Mongolia, recently discovered mineral deposits hold the potential to boost foreign direct investment well into the billions of dollars (Edwards 2013). In neighbouring Kazakhstan and the nearby Kyrgyz Republic, mining leads foreign direct investment, encouraging stronger connections to the international community than ever before (Deloitte 2013). In the developing countries home to the necessary mix of geology and climate, the figures estimating potential mineral wealth are staggering, but these figures come at a cost.

'Blood diamonds' famously fuelled civil war in Angola (Orogon 2004). The Democratic Republic of Congo's estimated US$24 trillion in untapped mineral wealth has not catalysed health and development (Morgan 2009). Instead, it drives conflict and greed. The average Congolese lives only 47.8 years and the country ranks at the very bottom of global corruption indices (Transparency International 2013). For Mongolian herders, the cost of mineral wealth is measured in water and history. In an area which receives an average annual rainfall of 80mm per year, herders worry the country's largest proposed mine will soak up or contaminate this sparse resource. They worry that the presence of an open pit mine and all its heavy equipment accoutrements and safety exclusion zones will defeat their nomadic way of life and endanger their indigenous culture and livelihood (Oyu Tolgoi Watch et al. 2012). Theirs is a story heard too often.

Downstream of the Tolukuma Gold Mine in the steamy, mountainous jungles of Papua New Guinea, Fané Village is hidden from interlopers, and immune to the passage of time and all its changes. Fané consists of about six thatched huts, some larger than others, set widely apart on a large patch of cleared dirt. A community garden rests along a verdant slope behind the huts. On any given day, children will be busy building forts in the dirt, their chickens plucking after them, dogs napping lazily. Since 1994, and with the approval of the Papua New Guinean Government, the mine has dumped its tailings – the waste rock leftover when you extract the valuable minerals from an ore – into the local river (Macdonald 2004). While efficient and convenient, the environmental and social costs come at the price of a river's death.

The impact of mining in developing countries is more severe than that experienced in the first world (Auty 1993). A dearth of strong governance mechanisms, lack of accountability, unethical business behaviour, involvement of paramilitary or private security forces, and tolerance of corrupt practices contribute to environments in which the needs of local communities or environmental protection are often sublimated to economic gain (Bebbington et al. 2008). Such experiences are so severe and so common, they have been dubbed the 'resource curse' (Auty 1993).

While developing countries weather the bulk of mining's negative impacts, these effects occur to varying degrees wherever the process takes place. In first-world countries, mining's impacts may be felt disproportionately by those individuals and communities who are more likely to face the marginalization and vulnerabilities associated with indigeneity, lower socio-economic status or rural living. Although such communities are more likely than their developing country counterparts to enjoy the considerations and protections of legitimate governments, environmental protection regimes, social programmes, welfare supports and economic compensation, mining nevertheless precipitates social, environmental and economic challenges.

In remote Western Australia mining takes place on and around lands which the earth's longest known human inhabitants call home. Yet until Australian law recognized the rights of indigenous people in 1993, native Australians had little or no ability to protect their territories. Although the overturning of the *terra nullius* (literally 'blank earth') doctrine granted indigenous Australians the ability to claim 'Native Title', certain lands were already irreparably lost to mining. In other situations, like that of Barramundi Gap in Western Australia's World Heritage Kimberley Region, considerable compromises were agreed. When Rio Tinto discovered its Argyle Diamond Mine in the late 1970s, indigenous Australians had limited legal recourse and faced a mining industry preaching a homily of speedy economic gain for the national good (Harvey and Nish 2005). Thus, when certain representatives of the Aboriginal community signed an agreement resulting in the destruction of the ceremonial Barramundi Gap site, other traditional owners and local Aboriginal communities opposed to the decision were bereft of choice and remedy.

The Argyle site eventually benefited from one of the most progressive community agreements negotiated within Australia (Harvey and Nish 2005). The Argyle Diamond Mine Participation Agreement, registered in 1995 as an Indigenous Land Use Agreement under Australia's *Native Title Act*, addressed the failures which led to the loss of Barramundi Gap. Based on principles of mutual recognition, obligation and Aboriginal owners' active participation in decision-making, the Agreement protects heritage sites, offers employment and training programmes and business development opportunities, and ensures those mine employees and contractors engaging with the mine receive cultural awareness training, among other things (Nish and Bice 2012). Such intensive and responsive engagements by mining companies remain rare. Yet they evoke the possibilities and potential of a responsible mining industry.

Attention to responsibility in the global mining industry

The date that the global mining industry developed a collective conscience is debatable. But one has undeniably been emerging, lesson by hard lesson, since at least 1998. That year, the heads of ten major global miners converged at the Global Mining Initiative (GMI), aiming to craft an industry-wide response to an enlarging litany of complaints and concerns. This included a series of unfortunate events which concentrated worldwide attention on the industry, especially the leaching of toxic chemicals into the Papua New Guinean Fly River downstream of BHP Billiton's Ok Tedi mine (Banks and Ballard 1997; Deegan et al. 2002). The GMI meeting spawned the International Council on Mining and Metals (ICMM) and the Mining, Minerals and Sustainable Development (MMSD) project, which certain experts peg as the genesis of corporate social responsibility (CSR) in mining (Dashwood 2004). Others tightly link the emerging conscience of global miners with the spread of globalization in the late 1990s, describing it as a by-product of corporations going multinational (Smith 2008).

Whatever the spark that concentrated industry-wide attention on mining's effects and the duties of companies to prevent, mitigate and respond to those impacts, responsibility is a widely acknowledged component of the contemporary mining industry and of multinational corporations more generally (Bondy et al. 2012). An entire field of research is devoted to making a business case for socially and environmentally responsible business practice (Crane et al. 2008). Dogged scholars persist in their quest for CSR's Holy Grail: irrefutable quantitative evidence linking corporate social performance with corporate financial performance (for reviews of such attempts, see Orlitzky et al. 2003; Salzmann et al. 2005). One recent study estimates that mining-community conflict propelled by poor company behaviour generates losses of up to US$20 million (net present value) per week, depending upon project size (Davis and Franks 2011). Other approaches couch the rationale for 'good' corporate behaviour within risk management frameworks, appealing to the concerns of chief operating officers and governance boards (Orlitzky and Benjamin 2001). Rio Tinto has, in the past, appealed to 'enlightened self-interest' (Harvey and Nish 2005). Efforts to 'create shared value' by leveraging existing corporate skills to create 'social good' are also gaining purchase (Porter and Kramer 2011).

Attempts to make a business case for CSR in mining are important. But they foster a tendency to ignore the policies and activities being implemented even while a solid business case remains lacking. From the capital city boardroom to the remote mine site, major mining companies' policies and actions reflect a general acceptance of their social responsibilities. Today's mining companies invest billions in community investment and CSR-related programmes. Major multinationals like BHP Billiton and DeBeers annually earmark at least one percent of pre-tax profits in community investment, a figure which has become a global benchmark (BHP Billiton 2012; DeBeers 2012). The professionalization of community relations roles reflects their expanding importance in the eyes of companies (Kemp 2010). Community consultative committees are no longer radical but conventional. Each year, major miners dedicate hundreds, if not thousands, of staff hours to data collection and sustainability reporting efforts. ICMM members are required to produce annual Global Reporting Initiative (GRI) Sustainability reports to an A+ level, meaning they must respond to 90 performance indicators covering economic, environmental, labour, human rights, product responsibility, social and mining-specific concerns.[5] Many are signatories to the UN Global Compact (UNGC) – the world's foremost sustainability framework, based on ten universal principles – while an increasing number are subject to financial transparency measures (Wintour 2013).

Industry-wide voluntary frameworks reflect further efforts to enshrine social responsibility and sustainable development as part of how miners do business. A spate of international and mining industry standards and frameworks appeared in the early 2000s, echoing a broader trend of global civil regulation (Dashwood 2007; Vogel 2008). Documents including the ICMM '10 Principles for Sustainable Development', the OECD's 'Guidelines for Multinational Enterprises', the Initiative for Responsible Mining Assurance (IRMA), the Extractive Industries Transparency Initiative (EITI), and the GRI's Mining and Metals Sector Supplement punctuated a cultural turn in companies' approaches to CSR. Public commitment to these frameworks indicated corporate recognition of duties to shareholders and stakeholders beyond the financial viability of the company. Such a triple bottom line approach is now commonplace but social (and to some extent environmental) activities and reporting remain primarily voluntary.

The policies, activities and public commitments of major global miners illustrate a critical shift in the way leading companies approach their business. Growing concern with 'earning and maintaining a social licence to operate' (Thomson and Boutilier 2010) and efforts to work 'beyond compliance' (Gunningham et al. 2004) signal an industry reflecting upon its values and practice. Much progress has been made since the MMSD launched just over a decade ago. Yet the bulk of this work has been completed in parcels, with industry-led initiatives and scholarly inquiry concentrating on broad level principles or very particular areas of impact. This has been vital and necessary work. It is now time to take stock of these efforts and to draw them together.

Five principles for responsible mining

Perhaps for the first time in the history of the mining industry, the foundations are in place to support responsible mining. Major companies openly recognize certain duties to the societies in which they operate, beyond financial viability. Governments institute regulatory regimes to protect and preserve environments, communities and human rights. Intergovernmental agencies provide global governance frameworks and accountability measures. Mine-affected communities access global networks and communications technologies to make their voices heard. This is a potentially transformational moment for a dominant global industry. One which demands articulation of a framework to support future practice.

The five principles of responsible mining offer one such framework. These principles include: **holistic assessment, community-based agreement-making, ethical decision-making, appropriate boundaries** and **good governance**. Responsible mining therefore brings together strong impact assessment methods and mitigation planning; principled relationships with local communities which prioritize equal community involvement in agreement-making; and sets boundaries for corporate activities and roles in states and communities within a framework of good governance. The chapters that follow journey across years and continents, through dusty deserts, over pot-holed roads, up mountains, down dense jungles, around board room tables, and into village squares to explore the pillars of responsible mining.

Chapter 2 provides a conceptual framework to guide our thinking about these issues. It explains the contemporary social processes institutionalizing CSR in the global mining industry. Along with Chapter 7, it is the most abstract of all the chapters in an otherwise pragmatically focused book, the 'lunchroom model' for CSR presented in Chapter 2 offers a chance to deepen understanding of just how particular concerns become incorporated into corporate considerations and to contemplate the implications of these processes within a broader social context. It delineates the key drivers which encourage the mining industry to adopt CSR and explores how these processes work in concert to perpetuate commitments to preventing and mitigating negative social and environmental impacts. These drivers include: progressive industry adoption of a language of values and responsibility to society and environment; widespread communication of such ideals; growing peer pressure between companies to behave responsibly; an ongoing quest to establish legitimacy with local communities; and the requirements of voluntary and state-based regulation. This understanding of the forces impelling the mining industry's concerns with social and environmental impacts provides a foundation on which to situate the five pillars of responsible mining.

Chapter 3 sets the stage for the discussion of the five principles to follow. Corporate policies, promises and activities are explored to illustrate global mining companies' current attitudes and behaviours toward their non-financial responsibilities. The 'great adoption' of CSR is explored through a comprehensive document analysis, enriched by interviews with mining executives, senior managers, community relations staff and those working closely with the industry. The chapter takes a deep dive into the public face of mining companies' approaches to social, environmental, governance, labour and human rights concerns through a critical analysis of a decade of sustainability reports published by five multinational miners between 2004 and 2013.

Chapters 4 through 7 draw upon a decade of research into the global mining industry. This includes extensive document analysis of mining company policies and publications, sustainability reports, government regulations and international frameworks. The information presented is immersed in current academic literature covering CSR, sustainable development, mining and extractive industries and organizational sociology. Statements from in-depth interviews completed between 2009 and 2015 are peppered throughout the chapters, adding voice and perspective to the ideas advanced (see the appendices for details of the research method and analysis). Observations from fieldwork undertaken at mine sites throughout Australia, Papua New Guinea and West Africa are included, as are those from conferences and meetings of industry, regulatory and international bodies working in the sector. While the research presented was executed for a variety of projects at different institutions, all were completed in line with relevant ethical research codes.

Chapter 4 introduces the first two pillars of responsible mining: holistic assessment and community-based agreement-making. Mining companies face requirements (to varying degrees) to test and predict their likely and possible social and environmental impacts wherever they operate. In most places, these requirements come in the form of social and environmental impact assessments. This chapter investigates and critiques the processes mining companies commonly use to formally identify their impacts in the first instance. It discusses the history of impact assessment and presents advances in concepts and methodologies which support more holistic impact assessments. The evidence and engagements possible through such good assessment practice offer a strong starting point for responsible mining.

Community-based agreement-making circumscribes debates about assessments and use of findings and aims to balance the costs of projects with desired community benefits. The chapter explores how community-based agreement-making incorporates local knowledge and concerns while providing community members with feedback about and responses to ongoing impacts identified through assessments and otherwise. Such processes can result in 'community impact and benefit agreements' which aim to secure the significant, continuing involvement of communities in determining their futures, relative to the mining development process.

For mining to be responsible, companies must engage meaningfully and ethically with communities. Chapter 5 explores this third pillar of responsible mining. Community relations practitioners tread difficult terrain in their efforts to understand and attend to community needs and concerns while prioritizing the viability of the mining operation. Ethical perspectives are introduced and discussed as a means of addressing the myriad challenges faced by those working at the coal face of mines and communities. Through cultural sensitivity, principled decision-making and acknowledging power differences between company representatives and community members, ethical relations are possible.

Ethical community relations underpin responsible mining's commitments to holistic assessment and community-based agreement-making. Principles of stakeholder engagement shape the conversation about the degrees to which mining companies should involve local, affected communities in decisions. Although ethics are often thought of as abstract, the chapter discusses leading approaches to ethical decision-making and suggests how these are well-aligned to the skills and expertise necessary for successful mining operations.

Appropriate roles and duties are also crucial factors in implementing and supporting responsible mining. Chapter 6 scours the boundaries between mining companies and government to investigate what happens when miners step in to provide infrastructure or services, such as roads or healthcare, which would otherwise rest with state authorities. Is it possible to establish sustainable communities when dependence is built upon a necessarily finite industry? Especially for developing countries, a balance must be struck between company provision of infrastructure and services and the potential power imbalances, paternalism and dependencies which may result. Appropriate boundaries between mining companies and governments, especially in relation to provision of infrastructure and services, form the fourth pillar of responsible mining.

For an agenda of responsible mining to succeed, strong site-, state- and country-level work must be ensconced within a legitimate framework of good governance. The chapter therefore goes on to investigate the origins and adoption of major international human rights and CSR initiatives, and evaluates their importance and usefulness for an accountable mining industry. The chapter analyses how such measures are being implemented and whether they are generating the changed attitudes and behaviours they espouse. Good governance provides a crucial fifth pillar and a guidepost for the continued reform, monitoring and answerability of the global mining industry to the communities it impacts.

Holistic assessment; ethical decision-making; community-based agreement-making; appropriate boundaries and good governance hold the potential to transform the mining industry. Yet theory alone yields limited results. Chapter 7 reviews the five pillars to establish theoretical propositions for how they can be better incorporated into corporate practice. CSR programmes can target research-identified community needs. Companies can provide further support and training for community relations professionals, especially in relation to ethics and awareness of company-community power imbalances. New means of assessing social impacts can be deployed and agreed criteria for a 'social licence to operate' established. Corporate-government boundaries can be considered more carefully and accountability can be boosted through more accessible, public disclosure of the mining sector's shortcomings and achievements.

The journey toward responsible mining has been decades in the making, and will continue for decades to come. It requires a belief in possibilities, a willingness to question the status quo, ingenuity and tenacity. Whether you've been suiting up in PPE since university or whether you thought PPE was that notoriously bad-tripping drug from the1960s,[6] there is much to be learned. For the mining company executive, the community relations practitioner, the plant engineer, this book is an opportunity to peer inside your own industry from a different perspective; to reflect on what's going well and what can be improved to leave a more positive legacy for mining. For community members living alongside open pits or underground channels, this is an exploration of what communities are experiencing, what companies are doing and how change is being achieved. For the NGO agents, consultants, researchers and social impact practitioners, the ideas within are a meditation on best practice to catalyse new and improved approaches and engagements. For everyone else, this book offers a chance to understand better an industry whose imprint and assets are enormous but through which it may be possible to effect substantial change.

Notes

1 By weight. The Eiffel Tower weighs in at an estimated 10,000 t.

2 For the record, the Forbidden City is about 720,000 m^2.

3 In 2012, Bahrain ranked at number 110 among all countries globally with a purchasing parity GDP of US$33.63 billion. One hundred and eighteen countries rank below it.

4 In a true story more plausible as a Tom Clancy novel, Lockheed Martin's 58,000 km^2 seabed mine area between the shores of Mexico and Hawaii was originally identified as mineral rich during the depths of the Cold War when the firm used a ship owned by reclusive American tycoon Howard Hughes to search for a sunken, nuclear missile-laden, Soviet submarine.

5 As of GRI version 3.1. The fourth version of the GRI, G4, was released in June 2013 with a two year grace period for reporters to make the transition. Few G4 reports had been released at time of writing, and the application levels (A, B, C) have been dropped from this latest version.

6 Rest assured, it's not. Boringly, it's personal protective equipment.

References

Ablett, J., A. Baijal, E. Beinhocker, A. Bose, D. Farrell, U. Gersc, E. Greenberg, S. Gupta and S. Gupta. 2007. 'The "Bird of Gold": The Rise of India's Consumer Market.' San Francisco: McKinsey Global Institute, 194.

Australian Bureau of Statistics. 2010. 'Labour Force Survey.' Canberra: Australian Bureau of Statistics. 6202.0.

Auty, R. M. 1993. *Sustaining Development in Mineral Economies: The Resource Curse Thesis*. London, New York: Routledge.

Banks, G. and C. Ballard, eds. 1997. *The Ok Tedi Settlement: Issues, Outcomes and Implications*. Canberra: National Centre for Development Studies and Resource Management in Asia-Pacific, Research School of Pacific and Asian Studies, Australian National University.

Bebbington, A., L. Hinojosa, D. Humphreys Bebbington, M. L. Burneo and X. Warnaars. 2008. 'Contention and Ambiguity: Mining and the Possibilities of Development.' *Development and Change* 39(6): 965–992.

BHP Billiton. 2012. 'We Value Sustainability: Sustainability Report 2012.' Melbourne: BHP Billiton, 50.

Bondy, K., J. Moon and D. Matten. 2012. 'An Institution of Corporate Social Responsibility (CSR) in Multinational Corporations (MNCs): Form and Implications.' *Journal of Business Ethics* 111(2): 281–299.

Brennan, E. 2013. 'The Next Oil? Rare Earth Minerals.' *The Diplomat*.

Cohen, M. 2013. 'Robotic Asteroid Prospector (RAP) Staged from L1: Start of the Deep Space Economy.' *NASA 2013 Spring Symposium*. Chicago: NASA.

Crane, A., A. McWilliams, D. Matten, J. Moon and D. Siegel, eds. (2008). 'The Corporate Social Responsibility Agenda.' *The Oxford handbook of corporate social responsibility*. Oxford: Oxford University Press, 3–18.

Dashwood, H. S. 2004. *Norms Dissemination and Corporate Social Responsibility: An Assessment of Global Processes and the Global Mining Initiative*. Montreal: International Studies Association annual meeting.

Dashwood, H. S. 2007. 'Towards Sustainable Mining: The Corporate Role in the Construction of Global Standards.' *Multinational Business Review* 15(1): 47–66.

Davis, R. and D. Franks. 2011. 'The Costs of Conflict with Local Communities in the Extractive Industry.' First International Seminar on Social Responsibility in Mining. Santiago: SR Mining, 13.

De Beers. 2012. 'Report to Society 2011.'

Deegan, C., M. Rankin and J. Tobin. 2002. 'An Examination of the Corporate Social and Environmental Disclosures of BHP from 1983–1997: A Test of Legitimacy Theory.' *Accounting, Auditing & Accountability Journal* 15(3): 312–343.

Deloitte. 2013. 'Doing business in Kazakhstan 2013: Reach, Relevance and Reliability.'

Edwards, T. and S. Paul. 2013. 'Mongolia Risk to Hurt Growth, Even with Oyu Tolgoi Start-Up, Election.' Reuters News Service.

Fangfang, L. 2013. 'Car Sales Back on Fast Track.' *China Daily*.

Geoscience Australia. 2013. 'Australian Atlas of Minerals, Resources, Mines and Processing Centres.' Canberra: Australian Government.

Gunningham, N., R. Kagan and D. Thornton. 2004. 'Social License and Environmental Protection: Why Businesses Go Beyond Compliance.' *Law and Social Inquiry* 29: 307–341.

Harvey, B. and S. Nish. 2005. 'Rio Tinto and Indigenous Community Agreement Making in Australia.' *Journal of Energy and Natural Resources Law* 23(4): 499–510.

Haxel, B., B. Hedrick and G. Orris. 2002. 'Rare Earth Elements: Critical Resources for High Technology.' Washington, DC: United States Geological Survey. 087-02. http://pubs.usgs.gov/fs/2002/fs087-02/ (accessed 2 March 2016).

International Council on Mining and Metals. 2012. 'Trends in the Mining and Metals Industry: Mining's Contribution to Sustainable Development.' London: ICMM.

International Energy Agency. 2012. 'Medium Term Coal Market Report 2012 Fact Sheet.' United States: International Energy Agency.

Kemp, D. 2010. 'Community Relations in the Global Mining Industry: Exploring the Internal Dimensions of Externally Oriented Work.' *Corporate Social Responsibility and Environmental Management* 17: 1–14.

Kosich, D. 2010. 'Officials Believe One of the World's Largest Gold Resources Lies Within BC.'www.mining.com/officials-believe-one-of-the-largest-gold-resources-in-the-world-lies-within-bc/ (accessed 2 March 2016).

Macdonald, I. 2004. 'Mining Ombudsman Case Report: Tolukuma Gold Mine.' Melbourne: Oxfam Australia.

Minerals Council of Australia. 2012. 'Australian Mining: This is Our Story.' Canberra: Minerals Council of Australia.

Morgan, M. J. 2009. 'DR Congo's $24 Trillion Fortune.' *African Business* 350: 52.

Moyer, M. and C. Storrs. 2010. 'How Much is Left? The Limits of Earth's Resources.' *New Scientist*.

Mudd, G. 2009. 'Historical Trends in Base Metal Mining: Backcasting to Understand the Sustainability of Mining.' Proceedings of the 48th Annual Conference of Metallurgists. Ontario: Canadian Metallurgical Society.

Nish, S. and S. Bice. 2012. 'Community Based Agreement-Making with Land Connected Peoples.' *New Directions in Social Impact Assessment: Conceptual and Methodological Advances*. F. Vanclay and A. M. Esteves, eds. Cheltenham: Edward Elgar, 59–78.

OECD. 2012. 'The Iron Ore Market in 2011.' Paris: OECD.

OECD. 2013. 'OECD Economic Outlook.' Paris: OECD, 1.

Orlitzky, M. and J. D. Benjamin. 2001. 'Corporate Social Performance and Firm Risk: A Meta-Analytic Review.' *Business Society* 40(4): 369–396.

Orlitzky, M., F. L. Schmidt and S. L. Rynes. 2003. 'Corporate Social and Financial Performance: A Meta-Analysis.' *Organization Studies* 24(3): 403–441.

Orogon, P. 2004. 'Blood Diamonds and Africa's Armed Conflicts in the Post Cold-War Era.' *World Affairs* 166(3): 151–161.

Oyu Tolgoi Watch, Bank Information Center, CEE Bankwatch, London Mining Network, Accountability Counsel and Urgewald. 2012. 'Review of the Oyu Tolgoi Copper/Gold Mine Environmental and Social Impact Assessment.' Ulaanbaatar, Mongolia: Mine Watch.

Porter, M. and M. Kramer. 2011. 'Creating Shared Value.' *Harvard Business Review*: January–February.

PricewaterhouseCoopers. 2012. 'Mine: The Growing Disconnect.'

Roche, C. and S. Bice. 2014. 'Anticipating Social and Community Impacts of Deep Sea Mining.' *Deep Sea Minerals and the Green Economy*. E. Baker and Y. Beaudoin, eds. Sydney: United Nations Environment Programme, 2.

Salzmann, O., A. Ionescu-Somers and U. Steger. 2005. 'The Business Case for Corporate Sustainability: Literature Review and Research Options.' *European Management Journal* 23(1): 27–36.

Smith, G. A. 2008. 'An Introduction to Corporate Social Responsibility in the Extractive Industries.' *Yale Human Rights and Development Law Journal* 11: 1–7.

Sonter, M. J. 1997. 'The Technical and Economic Feasibility of Mining the Near Earth Asteroids.' *Acta Astronautica* 41(4–10): 637–647.

Statistics Canada. 2013. 'Gross Domestic Product at Basic Prices by Industry (Monthly).' Quebec: Statistics Canada.

Thomson, I. and R. Boutilier. 2010. 'The Social License to Operate.' *SME Mining Engineering Handbook*. P. Darling, ed. Littleton, CO: Society for Mining, Metallurgy and Exploration.
Transparency International. 2013. 'Country Profiles.' www.transparency.org/country#COD (accessed 2 March 2016).
Tse, P.-K. 2011. 'China's Rare Earth Industry.' Washington, DC: United States Geological Survey.
United States Environmental Protection Agency. 2012. 'Copper Mining and Production Wastes.' Washington, DC: EPA.
US Central Intelligence Agency. 2013. *The World Fact Book*. Washington, DC: CIA.
Vogel, D. 2008. 'Private global business regulation.' *Annual Review of Political Science* 11(1): 261–282.
Wintour, P. 2013. 'UK and France to join global anti-corruption initiative.' *The Guardian*.

2 A lunchroom model of corporate social responsibility

'Responsible mining' seems at different times and locations: a great ambition, an oxymoron, a noble cause, weasel words, the new way of doing business or a pipe dream. Responsible mining may be any and all of these things, but this book asserts that it is possible and necessary. Future mining must be responsible mining. The concept extends from the adoption of 'corporate social responsibility' (CSR) or 'sustainable development' (SD) by multinational miners, primarily over the past two decades. This 'great adoption' of CSR has resulted in environmental, social and governance (ESG) concerns being integrated into how major miners do business. From operation sites to websites, community relations, SD, occupational health and safety and good employee relations are front and centre. While the degree to which the great adoption is rhetorical or sincere remains hotly debated, it nevertheless marks an important shift in the industry.

In this chapter, we explore how CSR became so prominent in the sector, reviewing its broader history alongside the development of scholarship to investigate it. This context is vital to understanding the five pillars of responsible mining, as they rest upon a theoretical scaffold steeped in social science and bolstered by years of research into CSR and the roles of firms in society. This theory equips us with a toolkit for understanding the current state of play more critically and more deeply. And it began with a simple question: If we are yet to have an indisputable business case for CSR but global miners have indisputably adopted it, how can we explain this?

This chapter introduces a 'lunchroom model of CSR' that explains the social drivers behind the great adoption. In social science, we have a theory about innovation: entrepreneurs innovate. Adopters look to those entrepreneurs they think are more legitimate or successful. The peer group copies. Innovation becomes internalized. Everyone looks the same. Innovation dies down. Attention turns elsewhere. A gross oversimplification, but you get the idea. Understanding CSR's institutionalization is crucial to discerning why and how the great adoption occurred in global mining and begins to answer the question of why profit-focused multinationals are looking beyond the financial bottom line.

The contested history of CSR scholarship

Mining companies' CSR activities are relatively recent, with Western, public attitudes about whether and how for-profit firms should carry out such commitments shifting substantially since the mid-twentieth century. Were you reading this in the 1970s while comfortably ensconced in your hanging bubble chair, your toes curled into the shagpile, the words of a Nobel prizewinning economist might sum up your opinion about these do-gooder companies. Writing in the *New York Times Magazine*, Milton Friedman (1970) famously stated that companies practicing CSR were 'preaching pure and unadulterated socialism. Business*men* [emphasis added] who talk this way are unwitting puppets of the intellectual forces that have been undermining the basis of a free society these past decades.'

Almost a half-century later and Friedman's article remains one of the most quoted and perhaps most misunderstood opinions on CSR. Despite appearing as dated as a red leather jumpsuit, the article is far more nuanced than its usual crib notes – 'The business of business is business' – suggest. Friedman's core concerns centred on what he saw as the overt and concerning political nature of CSR which encouraged CEOs to base business decisions on socio-political, not market, mechanisms. His incisive questions demanded that firms consider whether taking on roles and responsibilities beyond their profit-making charges was an irresponsible action in and of itself. Scholars, corporate leaders and CSR practitioners have struggled under the weight of Friedman's argument ever since.

Friedman's contentions are important. Not just because they capture a particular moment in CSR's great adoption, but because they sparked an entire genre of enquiry. Literally thousands of pages now document researchers' attempts to respond to and combat Friedman's assertions; to make a 'business case' for CSR. A business case for CSR offers firms the ability to make demonstrable, measurable and meaningful connections between their actions within society and their financial performance (e.g. Orlitzky et al. 2003; Waddock and Graves 1997). While some disagreement over the criteria defining a business case continues, for CSR it ultimately boils down to an appeal to the financial bottom line (Hanlon 2008). This is not to say that all research in this area is necessarily a riposte to Friedman but that his early assertions continue to reverberate within research and practice today (Barry 2000).

A business case for CSR?

Preoccupation with making a business case for CSR persists, whatever the agitator. This is despite a general acceptance that CSR is an important component of contemporary business, especially for those firms operating in industries, like mining and extractives, known for their considerable ESG impacts. Since 2010, 93 percent of CEOs surveyed by the United Nations Global Compact (UNGC) say that acting in environmentally sustainable, socially responsible ways is vital to their firms' success (United Nations Global Compact and Accenture 2010 and

2013). Seventy-six percent of these CEOs believe that 'embedding sustainability into core business will drive revenue growth and new opportunities' (United Nations Global Compact and Accenture 2013, 11). Figures like these support scholars' assertions that CSR is now mainstream among multinationals (Bondy et al. 2012).

Major global mining companies epitomize contemporary CSR practice. These mega-firms now regularly allocate one percent of pre-tax profits to community investment programmes, a figure which can stretch into the hundreds of millions (Bice 2013). In 2013 BHP Billiton, one of the world's largest commodities producers, invested US$245.8 million into community programmes globally (BHP Billiton 2014). Activities included biodiversity programmes in the Australian bush, donations to the Australian Indigenous Education Foundation, coastal land conservation in Chile, children's education programmes in Colombia and Pakistan, and eco-tourism support in Trinidad and Tobago. Colorado-based gold mining major Newmont delivered infrastructure, environmental and community development programmes and purchased over US$1.8 billion in goods and services from local suppliers (Newmont Mining Corporation 2014). Rio Tinto spent US$331 million on almost 2,200 community programmes in the same year (Rio Tinto Ltd. 2014), with almost one-third of funds allocated to education. The figures are staggering. The programmes are wide reaching. But such actions also raise important questions about the roles of corporations and governments, a topic to which we will turn in later chapters.

Progressive regulation now incorporates notions of CSR. The global financial crisis focused the minds of many governments on the need to enshrine more fulsome corporate responsibilities in legislation. Transparency measures inherent in the Dodd-Frank Act in the United States (US Congress 2010), the Law of the Right to Prior Consultation of Indigenous and Tribal Peoples in Peru (2010), and federal regulation for Australia's emerging coal seam gas industry (COAG Standing Council on Energy and Resources 2012) all speak to the acceptance of social responsibility as a vital component of contemporary extractive business.

Despite the variety of actions that indicate CSR's widespread acceptance, the business case remains inconclusive. Ambiguity around CSR's effectiveness sustains quantitatively geared studies seeking to show a link between corporate social performance (CSP) and corporate financial performance (CFP) (for a meta-analysis of these studies, see Orlitzky et al. 2003). Others explore the business-related drivers that stimulate firms' interest in CSR. University of California Berkeley Professor David Vogel (2005) argues that corporate concentration on CSR is fuelled by the proposition that socially responsible business practice bolsters the financial bottom. He wryly notes that it also forestalls regulatory red or 'green' tape (Vogel 2008). Anti-tape movements are visibly championed by the Australian mining industry (Australian Chamber of Commerce and Industry 2014; Davison 2014), by conservative US politicians (Rokita 2014) and by the Canadian Government's *Red Tape Reduction Plan* to 'identify irritants to business stemming from federal regulatory requirements' (Treasury Board of Canada Secretariat 2014). York University Schulich School of Business Professor Bryan Husted

(2005), meanwhile, adopts the language of 'risk management' to position CSR as a 'real option' for business leaders. He argues it allows corporate decision makers the opportunity to exercise environmental and social responsibilities where they result in favourable outcomes that also reduce business risk. Organizational sociology Professor Gerard Hanlon (2008) adds another dimension to the ways in which corporations approach CSR, arguing it is politicized and that the very nature of what we expect a corporation to be is shifting.

Business case approaches may also be more instrumental, viewing CSR as a means of generating competitive advantage, whether through 'shared value' creation (Porter and Kramer 2011), tapping frequently overlooked components of the supply chain (Prahalad 2003) or appealing to socially concerned investors (Kurtz 2008). Here, CSR may also be positioned alongside more traditional risk management (e.g. Husted 2005) or its potential influence on share value or stock desirability in the market (e.g. Bilbao-Terol et al. 2013; Kurtz 2008).

Although a definitive business case for CSR has yet to be made, studies focused on the CSP/CFP link contribute greatly to our understanding of how it affects business. This research provides insights into corporate governance and thinking, unveils the role of CEOs as major drivers of social responsibility, and investigates the relationship between a firm's available resources and its level of attention to CSR (for surveys of these contributions, see: Orlitzky et al. 2003; Salzmann et al. 2005; Kurucz et al. 2008). Perhaps most importantly, business case studies show that the ways in which CSR is operationalized appreciably influence any positive correlation between CSP and CFP (Orlitzky et al. 2003). The business case quest also reveals the changing nature of corporations within the context of globalization, a secondary but not insubstantial benefit of this research area (Kurucz et al., 2008).

Social scientific theory may seem irrevocably distanced from daily business practice. Yet the figures show that the why and how of CSR is no small consideration for major miners. The persistence of the business case problem is shaped partly by continued disagreement over CSR's definition, and by lack of unifying CSR theory, limited measurement tools and underdeveloped empirical methods (see e.g. Crane et al. 2008; Jamali 2007). These conceptual and methodological challenges are considerable, and they have tangible consequences. A long-held focus on the business case paradox risks losing sight of what effects the CSR policies and programmes already being implemented by numerous companies are having on the communities they seek to help, the organizations which implement them, and on the industries in which those companies operate. If CSR is indeed institutionalized, at least among multinational corporations – as evidence now suggests – and if a business case remains lacking, what might be a different way to understand its adoption and perpetuation? This chapter addresses this very question.

What drives CSR: new institutionalism and social mechanisms

To recap: a business case for CSR remains unproven. Yet firms continue to adopt and perform responsibilities beyond their financial duties to shareholders. We must then ask: if a business case cannot explain contemporary adoption and perpetuation of CSR, what else might? Fortunately, for over a century sociologists have been building theories that can assist us to address this question. Our discussion is about to get theoretical – but hang in there – it's worth it.

Old and new institutionalism

In order to understand the theory of CSR underpinning this book, it is helpful to take a moment to review the historical development of the social scientific study of organizations and institutions, of which CSR is one. Organizations and institutions, jointly and independently, have been a prime subject of sociologists, political scientists, economists and other scholars for over a century (Powell and DiMaggio 1991b; Scott 2001; Swedberg 2003). Famous for his theorizations of rationality and bureaucracy, Max Weber's work incited a renaissance in the study of institutions (Meyer and Rowan 1977; Scott 2001), generating a 'scholarly field' of organization studies (Scott, 1998). Ironically, however, few early institutionalist studies considered organizations (Scott 2001). It was only in the 1950s (Swedberg 2003) that analysis of institutions and organizations intermingled.

By the late 1970s, spurred largely by the work of Stanford Professor John Meyer, a 'new institutionalism' emerged through the publication of a series of theoretically allied, but not necessarily unified, studies (Powell and DiMaggio 1991b). Among these seminal works, Meyer's 1977 publications in *The American Journal of Sociology* (Powell and DiMaggio 1991b) proved fundamental. In the second of these two articles, Meyer and co-author Brian Rowan (1977) set out a new institutionalist research agenda, arguing that previous organizational theories were detached from the reality of the organization. While prior theories provided some insight into the formalities of bureaucracy, Meyer and Rowan suggested they were hamstrung by their inability or unwillingness to acknowledge the gap between formal organizational structure (bureaucratization) and the organization's often more informal, routine operation. They argued innovatively about organizational legitimacy by acknowledging that social forces affect organizations' credibility in society. This proposition was quite different to earlier organization theories which posited legitimacy as a given, based on norms of rationality.

Although the scholarly discussions of the late 1970s sparked substantial work in new institutionalism, it did not result in a unified approach (Powell and DiMaggio 1991b). Instead, three major new institutionalist paradigms developed: rational choice, historical and organizational (or sociological) institutionalism, with a fourth, discursive institutionalism, more recently acknowledged by certain scholars (Campbell 2004; Schmidt 2008). The details of these various paradigms are not so important, but understanding that new institutionalism may be applied in different ways via diverse perspectives is. Single-paradigm studies have made inroads into understanding such sundry topics as: societal structures and roles of the not-for-profit sector (Clemens and Powell 1998), consumer choice and boutique markets (Carroll and Swaminathan 2000), and the survival of organizations relative to their perceived managerial and technical legitimacy (Reuf and Scott 1998).

New institutionalist approaches are also used across a variety of academic disciplines, including economic sociology (see Dobbin 2004), political science (Hall and Taylor 1996) and international relations (Thomas 2004). These multidisciplinary studies have contributed important insights into the sociology of markets (Fligstein and Dauter 2007), competitive positioning of organizations within networks (Burt 1992; Granovetter 1973) and the rise of neoliberal economic structures (Campbell and Pederson 2001) and global phenomena (Campbell 2004).

Not only do scholars adopt different viewpoints to tackle the study of institutions, they continue to disagree over what, exactly, an institution is (Campbell 2004; Powell and DiMaggio 1991b; Zucker 1977). Although debate continues, it is now generally conceded that institutions are sociological phenomena, shaping and shaped by social context (Powell and DiMaggio 1991b). In other words, no single social actor creates or maintains an institution. It is the interaction of different actors within a given context that matters. The way in which 'institution' is defined and understood is strongly influenced by the paradigm or discipline from which it is being examined. Institutions are variously defined as: frameworks of rules; proscriptive actions; patterns of repetitive interactions; customs; governance structures; social arrangements that minimize transaction costs; sets of norms, rules or principles; or directly or indirectly agreed roles combined with conventions (Powell and DiMaggio 1991a). Importantly, throughout these various definitions, institutions are usually not organizations or vice versa. For the purposes of our discussion, an institution is defined as:

> A social order or pattern (Jepperson 1991) that is embedded in cultural and historical frameworks (Meyer and Rowan 1977), is shaped by and shapes cultural norms, but which is not necessarily a product of individuals setting an agenda (Campbell 2004) or of conscious design (Powell and DiMaggio 1991a).

From this definition, institutions are social patterns moulded by the widely accepted values and ideas (cultural norms) that are reinforced through their regular repetition. In this definition, institutions are made up of various 'social mechanisms', each of which contributes an essential working part to the amalgamated whole (Hedström and Swedberg 1998). Much like the contributions of cogs and wheels to a clock's inner workings, mechanisms' ultimate effects are not inherent in any one part (Davis and Marquis 2005). We can examine each mechanism independently – we can identify a fly pinion here, an hour wheel there – but it is not until they are put together that they become functional.

Defining CSR

Thinking about CSR in terms of the processes and patterns that shape it moves away from its more common conceptualizations, which usually focus on static, prescribed values frameworks, minimum standards, behaviour or outcomes. An institutional understanding of CSR looks instead towards dynamic social processes, hypernorms and attitudes. In this approach, it is not so much the outcomes of CSR that are important, but the social mechanisms that drive its adoption and perpetuation.

CSR today is generally well understood as positive corporate attention to non-financial ESG issues. But to generate a deeper understanding, we require a more sophisticated, conceptual definition. Drawing on our work above to define institutions, CSR as an institution can be defined as:

> A pattern of policies and activities undertaken by companies which are directly or indirectly related to their primary operations, but which are significantly influenced by social norms and expectations – as understood by companies and their stakeholders – concerning the company's social, environmental and economic behaviours and impacts.

If we think about CSR in this way – as a dynamic pattern of activities, as opposed to one-offs – important questions are raised. What are these patterns that tell us something is CSR? How are they created? And how do they reinforce one another? In theoretical terms, what does CSR as an institution look like?

For an institution like CSR to be formed, it undergoes a process of 'institutionalization'. Although bordering on the tautological, it is important to differentiate these terms. An institution is the social pattern that develops as a result of disparate social mechanisms working in concert. Institutionalization, following Ronald Jepperson (1991, 145), is, 'The process through which such a social pattern has attained the status of institution.' It is at this process that we will look to attempt to answer our questions about CSR.

We don't need a universal theory: striking a balance with mid-range theory

When sociologists attempt to answer theoretical questions like the ones raised in this chapter, we have to make some important choices about how we will build our theory; what we would like it to be capable of and what we believe it can do. CSR lacks a universal definition, and its contingency upon local contexts means that a universal theory – one that could be applied any place or time to explain it – is improbable and would fail to capture this nuance.

In the following sections we will attempt, instead, to build a 'middle-range' theory of CSR; one that strikes a balance between being universally descriptive while allowing for local diversity. Middle-range theory can open doors to more meaningful practical knowledge. In the words of theorists Joshua Margolis and James P. Walsh, '[It can] identify and probe the set of objectives, duties and concerns that arise when business organizations confront the question of whether to help redress human misery' (2003, 292). Peter Hedström and Richard Swedberg (cited in Davis and Marquis 2005, 336) suggest that middle-range theory '[offers] an intermediary level of analysis in-between pure description and storytelling, on the one hand, and universal social laws, on the other.' Pitched at a more empirical level, middle-range theory often examines social mechanisms, revealing the internal workings – cultural values, social norms, politics – that make up institutions (Campbell and Pederson 2001).

To build this 'middle-range' theory, we will look at CSR as it operates at three levels in the mining industry: the organizational field (industry), the formal organization (corporate headquarter) and the informal organization (operation site). In their study of institutional tensions and contradictions, Roger Friedland and Robert Alford (1991) outline the building blocks of 'an adequate social theory.' They argue a theory must analyse these three levels to achieve adequacy, from the micro to the macro. Such theory must encompass, 'individuals competing and negotiating, organizations in conflict and coordination, and institutions in contradiction and interdependency' (Friedland and Alford 1991). A multi-level analysis not only facilitates theory building, but also acknowledges the nuances and consequences of each level for our understanding of related institutions as a whole (Hannan and Freeman 1977).[1] These considerations are central to our quest to produce a more comprehensive understanding of responsible mining. They allow us to analyse how responsibilities are operationalized at industry (organizational field), corporate headquarter (formal organization) and operation site (informal organization) levels.

It is worth taking a moment here to define these levels of analysis, as we will refer to them throughout the chapters that follow. The industry level, or organizational field, is where many official policies and strategies are set out and benchmarks for public commitments to CSR are established. Organizational institutionalist gurus Paul DiMaggio and Woody Powell (1991) define an organizational field as:

> Those organizations that, in the aggregate, constitute a recognized area of institutional life: key suppliers, resource and product consumers, regulatory agencies, and other organizations that produce similar services or products.

Professor Dick Scott (2001) sees 'organizational field' as building on conventional conceptualizations of 'industry', which he defines simply as organizations which are grouped together based on their operating in 'domains' delineated by similar products or services. An organizational field may expand the concept of industry to include related organizations in the supply chain, regulators, governments, and the like, all of which occur within 'a recognized area of institutional life' (Scott 2001, 84). Analyses performed at the level of organizational field allow us to account for the contexts in which organizations operate by identifying a network of actors. For the lunchroom model of CSR set out below, it is the interactions and the relationships of actors in the network which are especially important, as they look to one another as benchmarks for their own behavioural choices and activities (Davis and Marquis 2005).

At the 'formal organizational' level, we will examine how individual mining companies define and implement CSR at their corporate headquarters. It is at this formal organizational level where management approaches are defined and CSR-related roles become professionalized and embedded within corporate structure, and through which CSR is connected to broader corporate strategy. In new institutionalist terms, an organization is constructed through rationalized rules and normative social pressures to create a structured and continuing organizational identity (Meyer and Rowan 1977). Think of any major brand whose logo and objectives make it immediately identifiable: Apple/innovation; Google/ information; Nike/sport; Walt Disney/magic. In all instances in our discussion, the formal organizational level represents parent company headquarters, which are the official site at which policy decisions are made, management approaches are delineated and commitments to CSR are determined. Mining companies' adoption of CSR is guided by shared beliefs and new institutionalists would say that these manifest in the form of 'myths and ceremonies' (Meyer and Rowan 1977). In other words, corporate headquarter values, mission statements and policies are frequently ritualized, playing out quite separately at operation sites (informal organizational level) (Meyer and Rowan 1977).

The 'informal organizational' level permits us to explore what the institutionalization of CSR means for communities and companies at operation sites. It is at this micro level where the strategies, policies and management approaches espoused by industry and corporate headquarters are shaped and adapted for local contexts and where, arguably, their impacts are most keenly felt. At this local community level, we arrive at a site for study of direct 'institutional environments' (Marquis et al. 2007), where the institution of CSR plays out in organizations' and individuals' daily lives. For many mining companies, this direct institutional environment exists within the communities that are host to their sites of operation. Here, the rationalized myths of the formal organization intersect directly with the social contexts in which companies operate (Meyer and Rowan 1977). As Scott (1991) suggests, it is also at this level where the social norms and values of community members may be most influential, fed back into mining companies and reshaping their very structure.

In the chapters that follow, we will regularly unpack mining companies' CSR activities into these three levels: organizational field (industry), formal organization (corporate headquarter) and informal organization (operation site). This allows us to investigate the gaps between CSR values and policies and on-ground activities and helps us to consider the factors that may influence this gap.

Middle-range theory: helpful, not perfect

Like all theoretical approaches, middle-range theory has its limitations. While it can help us to anticipate what to expect in other, similar situations, it is not predictive (Davis and Marquis 2005). Middle-range theory can allow near certainty that agents will act, but it cannot determine *how* or in which way (Davis and Marquis 2005). It does not aim to offer universally 'true' theories (Thelen 1999; Davis and Marquis 2005), making it less ambitious than sociology's traditional aim to produce universal laws (Knight 2001). But because of this, it is also more practicable and useful (e.g. Hedström and Swedberg 1998; Knight 2001; Davis and Marquis 2005).

New institutionalism and the study of CSR

New institutionalism offers a conceptual framework for better explaining CSR's dynamism (Brammer et al. 2012), the different ways it manifests (Matten and Moon 2008) and its different areas of focus (Kang and Moon 2012). A limited but growing group of researchers are applying new institutionalism to CSR (Brammer et al. 2012) in recognition that the approach allows us to address different research questions while simultaneously creating a more comprehensive understanding. Researchers Krista Bondy, Jeremy Moon and Dirk Matten's (2012) recent work finds that CSR is institutionalized among multinational corporations. Hevina Dashwood's (2012) application of rational choice and historical institutionalism buoyed her expansive exploration of normative aspects of CSR in the global mining industry. Matten and Moon's (2004) earlier work applied new institutionalist theories to study the worldwide spread of CSR, especially in the United States and Europe. They asserted that a new institutionalist approach encourages us to consider context and highlights how stakeholders interact and rely upon one another. Ruth Aguilera and Gregory Jackson's (2003) work on corporate governance acknowledges the individual and social pressures which shape it. Other contributions explore the boundaries between business and society and their implications for governance (Brammer et al. 2012). Studies like Dima Jamali and Ben Neville's (2011) emphasize the importance of investigating CSR at multiple societal and organizational levels. Jamali's (2010) work also encourages exploration of gaps between policy and practice, and the ways CSR is spread throughout organizations.

Dartmouth Professor John Campbell's work on CSR (2006 and 2007) adopts a 'second movement' new institutionalist perspective – in which various paradigms are used – to address the issue of boundaries and responsibilities. This allows him to define CSR in terms of minimum behavioural benchmarks. According to Campbell, firms must adopt a 'do no harm' approach to their internal and external stakeholders; a threshold above which companies may be considered to be enacting CSR and below which firms fall into socially irresponsible behaviour. While such a threshold-based definition provides a very helpful baseline, it may also unintentionally support reactive forms of CSR through an implicit focus on outcomes. Even if unintentional, a focus on outcomes suggests that CSR is the sum of aggregate behaviours, rather than a process. This is an important distinction when it comes to unpacking the social mechanisms which advance CSR; a task to which we now turn our attention.

Social mechanisms: a lunchroom model of CSR

At this point, we know that a CSR business case remains elusive and that, despite this, CSR is institutionalized in global mining. We know that, as an institution, CSR represents a pattern of values and behaviours, and that it is made up of social mechanisms that operate at three main levels: industry, corporate headquarter and operation sites. So what are the social mechanisms that make up CSR in the mining industry? And how do they work? The answers to these questions may lie in a social experience familiar to most of us: the high school lunchroom.

The research detailed throughout this book suggests that CSR is institutionalized through the work of four main social mechanisms. These are: discourse, mimesis, normative learning, and coercion. Or, in more familiar terms: shared language, peer pressure, group identity and discipline. Just as no adolescent wants to stand, tray in hand, adrift in the lunchroom and seeking acceptance at the cool table, corporations look to one another for acceptance. Today's companies exist in an almost clique-like environment where image and perception are imperative to recruit and retain quality staff, and to position themselves in the community and among their competitors. The same social pressures that applied in the high school lunchroom extend to the boardroom. In the sections that follow, we will unpack each of these mechanisms to build a lunchroom model of CSR.

A note on 'ideal types'

When Max Weber was busy establishing the discipline of sociology, he helpfully articulated one of the major difficulties inherent in attempting to explain complex and capricious human systems. Theorists need to be able to set out clear ideas and apply steady concepts to facilitate analysis but the world is rarely clear and never static. 'Ideal types' – like the social mechanisms explored below – are his answer. Ideal types allow us to create and test concepts while acknowledging that they are knottier in real life. For example, although it is certain that the number of social mechanisms contributing to CSR is greater than the four detailed here, those we examine have been selected carefully. It is also important to remain mindful that the social mechanisms comprising CSR intersect and overlap. This often makes distinct mechanisms difficult to distinguish from one another in practice. For the purposes of theory building, we must rely on interpretation and a somewhat artificial (but ultimately helpful) disaggregation of social mechanisms in order to facilitate analysis and discussion. It is also important to keep in mind that social mechanisms operate and are affected by broader socio-cultural contexts. With this in mind, let's get to work.

Shared language: discourse

The social mechanism of shared language (discourse) refers not only to words, but to the style in which they are delivered, why they are used, how they are understood, the structures in which they are deployed and the agents who use them (Schmidt 2008). Discourse plays a major role in the ways in which institutions are translated in different contexts (Campbell and Pederson 2001). It is steeped within cultural, historical and substantive circumstances (Schmidt 2008). For CSR this means that discourse is perhaps the most visible mechanism through which it is institutionalized. Paradoxically, it also contributes to varied understandings of CSR, especially between industries and cultures (e.g. Chapple and Moon 2005).

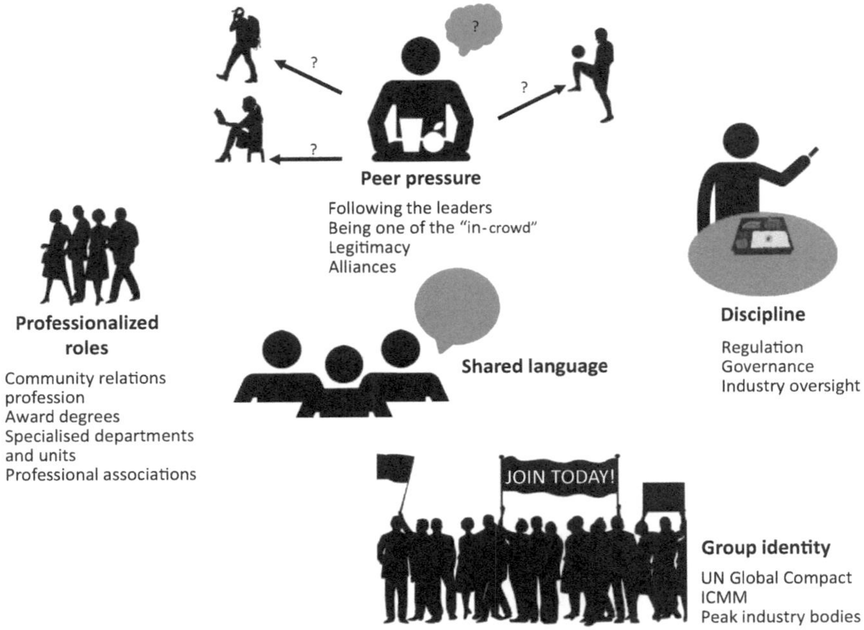

Figure 2.1 Drivers institutionalizing social performance

In the global mining industry, a shared language of CSR constructs it as a readily identifiable institution, even despite policies and management approaches which may differ between companies. This institutionalization occurs largely through the conscious and unconscious proliferation of language such as 'sustainable development', 'beyond compliance' (Hutchins et al. 2005) and 'social licence to operate' (Parsons and Moffat 2014). Across countries and operating environments, mining companies' CSR obligations are consistently referred to in terms of 'being a good corporate citizen' or 'social performance' (Yakovleva 2005).

In recent years the discourse of a social licence to operate has been particularly important in institutionalizing CSR in mining. It is prevalent at multiple social and organizational levels within the industry (Bice and Moffat 2014). It is deployed by peak industry bodies, such as the International Council on Mining and Metals (ICMM), which states a concern for companies' 'securing and maintaining a social licence to operate' (International Council on Mining and Metals 2010a, 96). Such statements echo industry-wide acceptance of CSR, acknowledging responsibilities beyond financial performance and regulatory compliance (see for example, International Council on Mining and Metals 2010b; Minerals Council of Australia 2005; Mining Minerals and Sustainable Development 2002).

Individual companies use the term to represent the importance they place on identifying and addressing their social impacts and consequent social changes (Joyce and Thomson 2000). This discourse is directly linked to mining companies' corporate social responsibility and is prevalent throughout major miners' public CSR communications (Bice 2014), especially sustainability reports (Parsons and Moffat 2014). Leading global miner BHP Billiton (2012, 2), for instance, opens a recent sustainability report by stating:

> Our ability to operate globally is dependent upon gaining access to natural resources and maintaining our licence to operate. SD is core to our business strategy; we integrate health, safety, environmental, social and economic factors into our decision-making'.[2]

This language posits CSR as central to community relations, mitigation of environmental damage and operational viability. Yet it is also a language steeped within mining's central concerns about natural resource exploitation, the historical controls of industry regulation, and an ultimate concern with business viability. As such, this shared language plays an important role in shaping how CSR is understood and communicated within and by the mining industry.

Discourse also signals the power structures inherent within institutions and their flexibility for institutional change (Schmidt 2008). In other words, the users, types and adaptability of language used about CSR play an important role in the concept's currency and relevance; its coolness. Were CSR unable to adapt linguistically to shifting stakeholder expectations and corporate developments, its lifespan would be limited. Moreover, discourse necessarily operates at various societal and organizational levels. From presenting broad socio-cultural implications for capitalist systems (Jones 1999) all the way down to informing or constraining individual identity (Foucault 1998), having a shared language is vital.

We have all experienced the spread of shared language – most people would know what an app is, something totally foreign only a few years ago – and it is through processes of diffusion and translation that new shared language promulgates. Diffusion represents the dissemination of largely static principles or practices through a related group of actors (Campbell 2004). For example, when such principles relate to human rights, diffusion may lead to creation of hypernorms – universal ethical standards (Wood et al. 2006). Translation, meanwhile, is the process through which globalized institutions are made sense of by local, regional and state actors (Campbell 2004). So, while CSR may play out differently according to context, translation tells us that it will always be identifiable but altered to suit the particular actors or organizations in question (Campbell 2004). Translation is strongly influenced by cultural diversity and also by diversity related to historical differences (Matten and Moon 2008).

Diffusion and translation represent the processes through which shared language is spread. But diffusion and translation cannot support unlimited proliferation. Instead they are bound by the national and international structures, regulations or agencies that promote CSR discourse. Where such support structures exist, socially responsible practices are more likely to be adopted and embedded (Jones 1999). As CSR discourse is shaped and reshaped through this 'structuration' (Giddens 1984) the institution itself is transformed. These processes are visible through widespread changes in CSR discourse over time – e.g. from corporate philanthropy to CSP, stakeholder management to corporate global citizenship (Carroll 2008). Discursive adaptations like these play an important role in how CSR is interpreted, adopted and implemented by organizations in different environments at different times.

The 'social licence' example is again illustrative here. This terminology has been deployed by global miners since the early 2000s (Joyce and Thomson 2000). Its widespread acceptance reveals an important, industry-wide shift from earlier language centred on 'health, safety and environment' towards a pseudo-regulatory licensing language (Bice 2014). In adopting such language, the mining industry's current conceptualization of CSR reflects contemporary global trends towards voluntary regulation (Vogel 2008) and stakeholder expectations about companies' responsibilities to communities (Prno and Slocombe 2012).

Discursive mechanisms also emphasize the 'normative' and 'cognitive schema' – regulated and internalized patterns – that shape a shared social reality (Scott 1991). Following Peter Berger and Thomas Luckman's (1967) seminal work on the sociology of knowledge, shared language plays a major role in this 'socially constructed reality' by shaping how we understand organizations and institutions (Meyer and Rowan 1977). Indeed, discourse often serves as the first contact point with any institution, and this holds true for CSR. Studies like that of Jamie Snider and colleagues (2003, 184) explore CSR web-based discourses in order to provide a 'gestalt of the ways in which the most successful firms globally describe their corporate social responsibility'. Much like the importance of understanding lunchroom slang to fitting in at school, discourse provides an entry point to understanding social phenomena.

The mining industry exemplifies this well. All major global miners now produce sustainability reports and devote portions of their webpages to 'sustainability', 'corporate responsibility' or 'communities'. In this way, CSR becomes a 'normalized schema', expected by stakeholders as a component of how mining companies do business with communities (Kemp 2010). While normative and cognitive schema around a discourse of CSR in global mining are apparent – the language is widely adopted and organized around particular, shared concerns – the depth to which such discourses are conceived is questionable. For instance, although landmark industry initiatives, such as the Extractive Industries Review (Salim 2004a and 2004b) and most mining companies' sustainability reports emphasize the importance of a social licence to CSR, little is done to define the criteria by which such a licence is granted. Even so, the discourse has become so powerful, it is now being leveraged by community members as a means of holding companies to account. In Australia's recent 'Upper Hunter Mining Dialogue' – an industry-led initiative to address the cumulative social and environmental impacts of the coal mining industry in a region of New South Wales also home to thoroughbred horse breeding, agriculture, vineyards and tourism – community members stated they were 'withholding' local companies' social licence to operate until such time that they felt their grievances had been clearly addressed (Australian Centre for Corporate Social Responsibility 2011). In theoretical terms, the shared language of CSR in the global mining industry rationalizes disparate terminology, propagates cognitive and normative schema, and offers a common platform for its analysis, reporting and regulation.

Peer pressure, group identity and discipline: isomorphic social mechanisms

The adoption of a shared discourse across the global mining industry is tightly linked to 'isomorphic' pressures institutionalizing CSR. Isomorphism is a kind of socially generated sameness, often visible in peer pressure and group identity. It results in similarities between organizations and their environments and between organizations and other organizations (DiMaggio and Powell 1991). Social mechanisms of mimesis, normative learning and coercion each contribute to isomorphism. This process results in institutions which are slow to change, not easily influenced by exogenous factors and which limit the behaviours considered appropriate for actors. Campbell's threshold definition of CSR, discussed previously, is reflective of this. As Powell and DiMaggio (1991, 11) describe it, 'Institutions do not just constrain options: they establish the very criteria by which people discover their preferences.'

To understand the social mechanisms contributing to isomorphism, it is critical to acknowledge that mining companies must not only appear rational individually, they must also coexist within a network or field, wherein structuration – the conditions through which social systems are maintained or altered (Giddens 1984) – creates and constrains opportunities for autonomy (Burt 1992) and legitimacy (Scott 1991). In other words, to be successful, mining companies must negotiate successfully within a social system where certain values and behaviours are deemed appropriate and acceptable by their peers.

In the case of CSR, isomorphic pressures driven by mimesis (peer pressure), normative learning (group identity) and coercion (discipline) are the basis for theorizations about global convergence of CSR policies and practices (Campbell 2005; Jamali and Neville 2011; Matten and Moon 2008). While new institutionalists differ in their opinions of whether structuration lends itself to diversity (Scott 1994) or homogenization (DiMaggio and Powell 1991), in the case of CSR (at least in the global mining industry), the latter appears to hold.

Peer pressure: mimesis

Mimesis – the modelling of actors on one another – is closely associated with uncertainty (DiMaggio and Powell 1991) and is therefore most influential during the early stages of CSR's development, much like the influence of peer pressure in early life stages. In the mining industry, CSR grew largely from a critical mass of international initiatives (Dashwood 2004; Schiavi and Solomon 2007) which were themselves responding to increased global media and public attention to mining-related impacts (Deegan et al. 2002). The historic development of CSR in mining can help us to understand the influence of mimetic processes on CSR's emergence in the industry.

Socio-environmental mining disasters in the 1980s spurred scrutiny of major miners, created a crisis of confidence and demanded response. Canadian author Havina Dashwood pinpoints 1998's Global Mining Initiative and subsequent ICMM, and the Mining, Minerals and Sustainable Development (MMSD) project as the originating source of CSR in mining (Dashwood 2004). Gare Smith (2008), writing in a special edition of the *Yale Human Rights and Development Law Journal* on 'Corporate social responsibility in the extractive industries', tightly links CSR's birth with the spread of globalization in the late 1990s, especially the spread of multinational enterprises. In these instances, the mining industry responded to increased pressure for socially and environmentally responsible behaviour, with major miners representing the early adopters (Dashwood 2012) and a growing cohort of middle miners modelling their CSR on other firms over time. These modelling processes, and the shift from major to middle miners' adoption, are visible, for instance, in the now widespread production of Global Reporting Initiative (GRI) sustainability reports. In 2013, the GRI recorded 179 sustainability reports produced in the mining industry, representing giants like BHP Billiton to middle miners like Indonesia's ANTAM (see Global Reporting Initiative 2014).

While mimetic processes can generate widespread adoption of norms and behaviours, they can also stymie innovation. Especially in the early stages of institutional development, adopters look to entrepreneurs whom they 'perceive to be more legitimate or successful' (DiMaggio and Powell 1991, 70). This habit of looking to those who have come before leads to convergence of values and practices, even across diverse international contexts. Mimesis, therefore, is closely linked to legitimacy; as new institutions form, peers look to one another to determine what is right and acceptable. As more actors adopt stances similar to their predecessors, particular values, behaviours and policies are affirmed. Convergence dominates.

Convergence is visible in the global mining industry where companies' CSR has been concentrated largely into a catalogue of agreed priorities and activities. The widespread adoption of common CSR practices by mining companies includes: local employment; procurement and shared infrastructure programmes; apprenticeships and training; donations and other charitable contributions to civil society groups, NGOs and other organizations; direct funding or delivery of welfare programmes; construction of civic infrastructure (e.g. hospitals, health clinics, schools, recreation centres, sports fields and non-employee housing); establishment and funding of company-controlled trusts, funds or foundations (i.e. local development NGOs); and support for 'outsourced' community development services, particularly those associated with national and international (supply-side) organizations (Harvey and Bice 2014).

The extent to which mimetic processes contribute to isomorphism in CSR practices in the global mining industry raises important questions about the status and effectiveness of CSR priorities and activities now and into the future. Emerging research is beginning to question the form and effectiveness of commonly deployed mining company approaches to CSR. From a new institutionalist perspective, the strong isomorphism visible in mining companies' and mining industry approaches to CSR suggest that the institution has advanced beyond the innovation stage, wherein early adopters were motivated by an entrepreneurial desire to improve performance. If this is the case – and recent research around a social licence to operate and related concerns about credibility would certainly seem to suggest this (Owen and Kemp 2013; Prno 2013) – then contemporary CSR in mining is better understood as 'normatively sanctioned', further supporting the argument for its institutionalization.

Group identity: normative learning

Normative learning also contributes substantially to CSR's institutionalization. Linked closely to professionalization, normative learning encompasses efforts to 'define conditions and methods of work' (DiMaggio and Powell 1991, 70), as well as related processes such as establishing industry member organizations. The institutionalization of CSR is also linked to its progressive ceremonialization in organizations' professions, programs, priorities and procedures. In other words, normative learning relates closely to the practices that help to establish a group identity. As Meyer and Rowan (1977, 344–345) predicted in the late 1970s:

> As the issues of safety and environmental pollution arise, and as relevant professions and programmes become institutionalized in laws, union ideologies, and public opinion, organizations incorporate these programs and professions.

The development of CSR in the global mining industry reflects such normative learning well. Over several decades mining companies' CSR has transformed from concentration on human resources practices and relatively limited public disclosure (Guthrie and Parker 1989) to comprehensive ESG programming backed by progressively professionalized staff (Kemp 2010). The establishment of the ICMM in 2001 created a member group comprised of the world's leading mining and metals companies 'to advance their commitment to sustainable development' (International Council on Mining and Metals 2014). This and other mining industry associations consciously and unconsciously assert isomorphic pressures concerning expectations for 'best practice' upon firms through guidelines and signatory commitments to global CSR frameworks, such as the UNGC or OECD Guidelines for Multinational Enterprises. Manifold other pressures systemically influence multinational mining companies towards homogeneity (Shapiro et al. 2007).

Other key components of this normalization include establishment of expected educational credentials and completion of particular degrees or training programmes. This is followed by similar role offerings across organizations, allowing for flows of personnel within the industry. For CSR, such normative processes can be seen in the offerings of top business schools (Symonds 2011), such as Harvard and Wharton (Harvard University 2011; The Aspen Institute: Centre for Business Education 2010), which teach 'corporate responsibility' within core leadership modules. Others, like the UK's Cranfield School of Management, offer students the opportunity to specialize in 'sustainable business' (Cranfield University School of Management 2011). Further academic resources are allocated to the study of business' contemporary role within society in places like Boston College's Center for Corporate Citizenship or the University of Queensland, Australia's Centre for Social Responsibility in Mining. Graduates of these and similar programmes enter a job market in which leading global companies now employ individuals to fill roles such as 'Manager, Environmental and Social Responsibility', 'Director, Corporate Social Responsibility', or 'Environment, Health and Safety Specialist' (see job listings on: BrighterPlanet.org, for example). Other recent studies suggest expertise in CSR will become 'a specific managerial competency' (Blowfield and Murray 2008). Growth of CSR roles and professionalization of the field exert isomorphic pressures further institutionalizing corporate social responsibility.

Discipline: coercion

Coercion contributes to isomorphism through 'formal and informal pressures exerted on organizations by other organizations upon which they are dependent and by cultural expectations in the society within which organizations function' (DiMaggio and Powell 1991). It is helpful to think of coercion as a form of discipline, especially one carried out via regulation. For the purposes of our discussion, regulation is not necessarily formal or codified in law. Instead, coercive regulatory mechanisms include formal codes and laws plus those voluntary guidelines, performance indicators, and principles or frameworks that set standards for CSR performance. Returning to our lunchroom, this style of isomorphism is most like the good behaviour that results from the presence of the teachers' table.

In practice 'voluntary' regulation is perhaps more strongly associated with CSR than its more formal counterpart (Eigen 2007). Voluntary regulation helps to define areas of concern; creates recognized minimum standards; supports a shared dialogue among organizations; suggests a particular global order; fosters accountability and competition through improved performance comparability; may seek to address deficient regulatory or governance standards (Aguilera and Cuervo-Cazurra 2004); may equally devolve or divert authority from the level of the state to the industry or organization (Vogel 2008); and allows space for the establishment of standards which may be more industry specific and, by extension, more rigorous than generic, formal regulation (Brunsson and Jacobsson 2000; Campbell 2007).

Much research concerning the regulation of CSR has dealt primarily with the debate over whether regulatory codes should be requisite, as opposed to voluntary (see for example, Eigen 2007; Hess 2007; Sadler 2009; Sethi 2003). A new institutionalist perspective on the regulation of CSR suggests that, regardless of whether regulation is legislated by the state or introduced by private or non-profit organizations, its proliferation and uptake by corporations is what matters (Campbell, 2007). As more and more companies choose to adopt particular principles or report against certain frameworks, the peer pressure to use those frameworks increases, diffusing values, discourse and expectations across industries and increasing coercive isomorphic pressures (Galaskiewicz 1991).

Ethicist Laura Hartman (2008) and her colleagues (2003) posit similar arguments when they suggest that profit-driven organizations must shift and adapt to globalized markets through bottom-of-pyramid approaches (Prahalad 2003) and adoption of ethical or socially responsible practices. Studies like that of Sükrü Özen and Fatma Küskü (2009) assert that, where industries are highly concentrated and competition is high – as in the global mining industry – voluntary regulation is more likely to be adopted as a means of legitimacy-seeking. Firms must act according to 'collectively valued purposes' in order to be legitimized within the marketplace (Meyer and Rowan 1977, 349). But even where firms adopt socially acceptable structures based on collective values and ideals, if these structures are internally focused or too novel, they may not provide the desired legitimation (DiMaggio and Powell 1991). Additionally, despite their claims to rationality, mythologized practices may not be the most efficient (Meyer and Rowan 1977).

Taken together, the above considerations help partly to explain the wide adoption of voluntary CSR regulation by mining companies. These factors also illustrate why practices are more isomorphic than diverse. First, although individual mining companies might be able to produce better tailored, thorough or robust auditing frameworks or performance indicators, these may not carry the necessary legitimacy. Initiatives such as the GRI provide an 'external criteria of worth' (Meyer and Rowan 1977, 350) for valuing CSR within firms. Secondly and consequently, companies employ frameworks that are necessarily broad (for instance, the GRI G3 comprised 79 generic performance indicators plus sector-specific supplements) and therefore somewhat inefficient. In this rather complex manner, a CSR practice which is valued externally but which conflicts with rational efficiencies, is established and perpetuated.

The uptake of voluntary regulation relative to CSR is also linked to its sheer volume of proliferation. In his 2008 meta-analysis, David Vogel cited approximately 300 'global civil regulation' codes intended for the voluntary regulation of CSR. In practice, the most common global voluntary regulations concerning CSR include the GRI and the UNGC (Global Reporting Initiative 2010; Global Reporting Initiative and United Nations Global Compact 2010). Frequently used together (Global Reporting Initiative and United Nations Global Compact 2006), these frameworks foster consensus around the key issues on which corporations seeking to act responsibly should focus. In addition to these broader frameworks, mining companies' CSR is largely informed and regulated by the ICMM Sustainable Development Principles, the Extractive Industries Transparency Initiative (EITI), IFC Social and Environmental Performance Standards, the Equator Principles and the recent UN Guiding Principles on Business and Human Rights. Other important frameworks include the OECD Guidelines for Multinational Enterprises, ISO 26000 and country or commodity-specific guidelines such as the Minerals Council of Australia's *Enduring Value Framework*, The Mining Association of Canada's *Towards Sustainable Mining* framework and the Kimberley Accord and related Process to prevent 'blood diamonds'. These frameworks construct a voluntary monitoring framework which both shapes and is shaped by (Power 1997) organizations' and stakeholders' expectations about what constitutes CSR.

Following Michael Power (1997), these frameworks can be viewed as auditing tools which represent 'institutionalized products' used for 'external legitimation'. Even in spite of their voluntary nature, these standards lend credibility to corporate social responsibility efforts while also disseminating a global CSR discourse, linking and homogenizing patterns of socially responsible behaviour through exertion of institutional pressure.

Discipline is also closely linked to societal expectations and the exercise of coercive social power, as opposed to regulatory authority (Scott 1991). For instance, we can see coercion via social pressures in the historical adoption of CSR in global mining discussed above. The responses of companies to community pressures illustrate the role and agency of stakeholders in influencing the adoption and perpetuation of CSR (Delmas and Toffel 2004). Community interest groups such as 'No Dirty Gold' and 'Lock the Gate' act as civil society watchdogs of the mining and extractives industry. Nongovernmental organizations, including Oxfam America and Australia's mining campaigns, Transparency International, Publish What You Pay and Save the Children, place the weight of their organizations behind social pressures for environmentally and socially responsible mining. Publications such as *Business for CSR*, *CSRWire*, *Ethical Corporation* and *Ethical Performance* now regularly monitor companies' social and environmental successes and failures. Intergovernmental agencies and international standards organizations, including the United Nations, World Bank, International Finance Corporation and International Organization for Standardization all play a vital role in encouraging minimum benchmarks for companies' CSR behaviour through codes and standards. Combined, these activities establish a consistency of approach, improve comparability between companies and industries, and build credibility and legitimacy.

Much as mining companies look to one another concerning shared language, activities and professionalization, the voluntary frameworks and standards they have widely adopted normalize and reinforce the issues, practices and reporting indicative of socially responsible firms. Regulation (even where it is not regulation by the state) creates a degree of stability, promoting isomorphism in which company representatives look to other firms' agendas, interests and behaviours when making decisions about their own actions or choices (Fligstein 1991).

Research concerning the roles of regulation and auditing offers further evidence of the disciplining pressures exerted through voluntary CSR frameworks and reporting initiatives. Such research suggests that frameworks, such as the GRI, must be understood as both reflecting and creating agendas (Power 1997). As Campbell (2004, 149) states, 'Many institutionalists agree that regulative institutions shape people's perceptions of their interests.' Similarly, Power (1997) argues that regulatory frameworks frequently have the unintended effect of shaping the very programmes they seek to measure and hold accountable. Like the timeless 'chicken or egg' question, it is impossible to determine with any certainty the extent to which frameworks like the GRI shape the focus of corporations' CSR activities, or the extent to which pre-existing CSR behaviours influence the frameworks themselves. Regardless of the direction of force, however, what is clear is that voluntary regulation in the form of reporting frameworks, societal pressures exercised by community and organizational stakeholders, and subsequent behaviours reinforce a cycle in which CSR is reified, legitimized and institutionalized.

The wider debate over the regulation of CSR is influenced by the context of neoliberalism (Levy and Kaplan 2008). For scholars like Peter Haas (2004, cited in Levy and Kaplan 2008), the rise of neoliberalist policies globally has fuelled regulatory concerns and fostered support for non-state, voluntary initiatives. In a somewhat ironic twist, the rise of neoliberalism appears to have been one of the factors most significant in the global diffusion of CSR (e.g. Levy and Kaplan 2008; Newell 2008; Sadler 2009). In practice, countries advancing neoliberalist agendas, like the US and the UK, have been hotbeds of CSR developments (Sadler 2009). Far from discouraging corporate social responsibility, neoliberalist states appear indirectly to have created space for the establishment of CSR marketplaces comprised of specialist consultancies, organizations, training providers and auditors (Newell 2008; Sadler 2009). This may be due to underlying ideological consistencies or, as David Levy and Rami Kaplan (2008) suggest, to the widespread situation where 'the CSR movement has avoided challenging the core economic structures and managerial prerogatives of contemporary market societies, which retain a high degree of legitimacy.' From a neoMarxist perspective, CSR is a necessary antidote to the spread of neoliberalism in a globalized market. Professor Peter Newell suggests, 'There is a need to match the globality of capital's reach with frameworks of social regulation that are up to the task' (2008).

Although coercive mechanisms play an important role in the institutionalization of CSR, they also hold within them potential threats to its perpetuation. It is important to be wary of 'juridification' – the potential of frameworks, standards, rules and the like to overwhelm organizations with minutiae to such an extent that compliance becomes onerous and, possibly undesirable (Haines and Sutton 2003). Contemporary CSR faces this regulatory dilemma. Although many companies choose to employ world-leading frameworks, such as the GRI and UNGC, an estimated 300 alternatives exist for voluntary CSR governance alone (Vogel 2008). Lack of formal regulation may lead to excessive choice (Haines and Sutton 2003) that can discourage firms' strategic pursuit of CSR. More formal regulation might be desirable, if for no other reason than to refine and reduce the number of guidelines corporations employ. As Professor Fiona Haines and the late Adam Sutton (2003) noted, 'excessive compliance demands can be noxious', a circumstance towards which CSR regulation may be heading and which must be avoided.

The difficulties of peer pressure: isomorphic challenges

The isomorphic mechanisms institutionalizing CSR in mining – peer pressure, group identity and discipline – present several classic problems for organizations. Perhaps foremost is the difficulty that results from the multi-source nature of 'rationalized myths' – those policies and practices informed by societal expectations that become 'ceremonialized' by firms to demonstrate their rationality to both internal and external actors (Meyer and Rowan 1977). Rationalized myths contribute to cleavages between formal organizational policies and structures and actual behaviours, a process referred to as 'decoupling'. In other words, in order to function under societal expectations, irrational firms create rationalized policies and ideals that, for various reasons, they are unable to institute but which suggest they are rational. Weber (1922/1946) analogized this denial of irrationality to an attempt to balance an equation. Although the resulting sum is uneven, the actor simply ignores the difference and assumes equilibrium to achieve the rational result she desires. The denial of difference between the rationality and the irrationality of organizational elements *mythologizes* them as rational and imbues them with distinct and important social mores (Meyer and Rowan 1977). Decoupling becomes the negative space between organizational rhetoric and practice. Rationalized myths about firms' responsibilities and contributions to society shape how the organization is understood but also carry conflicting 'ceremonial' rules.

Friedman (1970; *The Economist* 2005), for example, viewed CSR as diametrically opposed to the organization's rational aim of financial viability. Through isomorphism the two myths – CSR and profit-motive – collide, problematizing concerns with priorities and efficiency. Moreover, isomorphic pressure to adopt and perpetuate CSR is very often in conflict with the technical activities and demands for efficiency placed on the organization. Attempts to prioritize or implement CSR may conflict directly or be inconsistent with the ceremonial rules of production. These tensions in isomorphic pressure help to explain why corporations and scholars struggle to determine the business case for CSR. The conflict inherent in competing myths and isomorphic pressures makes it especially difficult to link ceremonial and efficiency requirements, contributes to decoupling between formal and informal organizational structure, and encourages a 'logic of confidence' wherein 'elaborate displays of confidence, satisfaction and good faith' may be made regardless of the actual situation (Meyer and Rowan 1977). Here, inspection and evaluation may also be minimized or ceremonialized, a situation evidenced in the prevalence of voluntary CSR regulation (Vogel 2008).

Conclusion

This chapter presented the theoretical scaffold on which the five principles detailed in the following chapters rest. We have explored the history and context of CSR – and the most common ways in which scholars have analysed it – to offer explanations for corporate attention to issues beyond the financial bottom line. A new institutionalist perspective offers a different and productive way of thinking about responsible mining. It allows us to conceptualize CSR as an institution, consequently recognizing its dynamism, its close relation to social processes, and its status as a pattern of particular beliefs and behaviours. We have developed a lunchroom model of CSR, unpacking the key social mechanisms – shared language (discourse), peer pressure (mimesis), group identity (normative learning) and discipline (coercion) – which help to explain its adoption and perpetuation.

Throughout our theoretical discussion illustrative examples from across the industry demonstrated the ways these social mechanisms work in concert to institutionalize CSR in global mining. CSR in mining – by any name – is no longer an added extra or a nice-to-have. It is essential and it undergirds responsible mining.

Notes

1 While Hannan and Freeman (1977) represented political ecology, their long-standing criticism that new institutionalists often fail to define their levels of analysis remains salient today.

2 We will explore the influence of shared language through mining companies' sustainability reports in depth in Chapter 3, when we look at a decade of mining company sustainability reporting.

References

Aguilera, R. and G. Jackson. 2003. 'The Cross-National Diversity of Corporate Governance: Dimensions and Determinants.' *Academy of Management Review* 28(3): 447–465.

Aguilera, R. V. and A. Cuervo-Cazurra. 2004. 'Codes of Good Governance Worldwide: What Is the Trigger?' *Organization Studies* 25(3): 417–446.

Australian Centre for Corporate Social Responsibility. 2011. 'Upper Hunter Mining Dialogue: Report on the Stakeholder Survey for the NSW Minerals Council.' Melbourne: Australian Centre for Corporate Social Responsibility, 40.

Australian Chamber of Commerce and Industry. 2014. 'Acci National Red Tape Survey 2013.' Canberra: ACCI, 20.

Barry, N. P. 2000. 'Controversy: Do Corporations Have Any Responsibility Beyond Making a Profit?' *Journal of Markets and Morality* 3(1): 100–107.

Berger, P. L. and T. Luckman. 1967. *The Social Construction of Reality: A Treatise in the Sociology of Knowledge*. London: Allen Lane.

BHP Billiton. 2012. 'We Value Sustainability: Sustainability Report 2012.' Melbourne: BHP Billiton, 50.

——. 2014. 'Our Contribution: BHP Billiton in the Community.' Melbourne: BHP Billiton, 50.

Bice, S. 2013. 'No More Sun Shades Please: Experiences of Corporate Social Responsibility in Remote Australian Mining Communities.' *Rural Society Journal* 22(2): 138–152.

——. 2014. 'What Gives You a Social Licence? An Exploration of the Social Licence to Operate in the Australian Mining Industry.' *Resources* 3(1): 62–80.

Bice, S. and K. Moffat. 2014. 'Social Licence to Operate and Impact Assessment.' *Impact Assessment and Project Appraisal* 32(4): 257–263.

Bilbao-Terol, A., M. Arenas-Parra, V. Canal-Fernandez and C. Bilbao-Terol. 2013. 'Selection of Socially Responsible Portfolios Using Hedonic Prices.' *Journal of Business Ethics* 115(3): 515–529.

Blowfield, M. and A. Murray. 2008. 'The Future of Corporate Responsibility.' *Corporate Responsibility: A Critical Introduction*. M. Blowfield and A. Murray, eds. Oxford: Oxford University Press, 361.

Bondy, K., J. Moon and D. Matten. 2012. 'An Institution of Corporate Social Responsibility (CSR) in Multi-National Corporations (MNCS): Form and Implications.' *Journal of Business Ethics* 111(2): 281–299.

Brammer, S., G. Jackson and D. Matten. 2012. 'Corporate Social Responsibility and Institutional Theory: New Perspectives on Private Governance.' *Socio-Economic Review* 10(1): 3–28.

Brunsson, N. and B. Jacobsson. 2000. *A World of Standards*. New York: Oxford University Press.

Burt, R. S. 1992. *Structural Holes: The Social Structure of Competition*. Cambridge: Harvard University Press.

Campbell, J. 2004. *Institutional Change and Globalization*. Princeton: Princeton University Press.

——. 2005. 'Why Would Corporations Behave in Socially Responsible Ways? An Institutional Theory of Corporate Social Responsibility.' *Academy of Management Review*: 1–45.

——. 2006. 'Institutional Analysis and the Paradox of Corporate Social Responsibility.' *American Behavioral Scientist* 49(7): 925–938.

——. 2007. 'Why Would Corporations Behave in Socially Responsible Ways? An Institutional Theory of Corporate Social Responsibility.' *Academy of Management Review* 32(3): 946–967.

Campbell, J. and O. Pederson. 2001. 'Introduction: The Rise of Neoliberalism and Institutional Analysis.' *The Rise of Neoliberalism and Institutional Analysis*. J. Campbell and O. Pederson, eds. Princeton: Princeton University Press, 1–24.

Carroll, A. B. 2008. 'A History of Corporate Social Responsibility: Concepts and Practices.' *The Oxford Handbook of Corporate Social Responsibility*. A. Crane, A. McWilliams, D. Matten, J. Moon and D. Siegel, eds. Oxford: Oxford University Press, 19–46.

Carroll, G. R. and A. Swaminathan. 2000. 'Why the Microbrewery Movement? Organizational Dynamics of Resource Partitioning in the U.S. Brewing Industry.' *American Journal of Sociology* 106(3): 715–762.

Chapple, W. and J. Moon. 2005. 'Corporate Social Responsibility (CSR) in Asia: A Seven-Country Study of CSR Web Site Reporting.' *Business and Society* 44: 415–443.

Clemens, E. and W. Powell. 1998. *Private Action and the Public Good*. New Haven: Yale University Press.

COAG Standing Council on Energy and Resources. 2012. 'The Draft National Harmonised Regulatory Framework: Coal Seam Gas.' Canberra: Council of Australian Governments (COAG), 80.

Crane, A., A. McWilliams, D. Matten, J. Moon and D. Siegel. 2008. 'The Corporate Social Responsibility Agenda.' *The Oxford Handbook of Corporate Social Responsibility*. A. Crane, A. McWilliams, D. Matten, J. Moon and D. Siegel, eds. Oxford: Oxford University Press, 3–18.

Cranfield University School of Management. 2011. 'Doughty Centre for Corporate Responsibility.' www.som.cranfield.ac.uk/som/p14340/Research/Research-Centres/Doughty-Centre-Home (accessed 2 March 2016).

Dashwood, H. S. 2004. 'Norms Dissemination and Corporate Social Responsibility: An Assessment of Global Processes and the Global Mining Initiative.' *International Studies Association annual meeting*. Montreal.

——. 2012. *The Rise of Global Corporate Social Responsibility: Mining and the Spread of Global Norms*. Cambridge: Cambridge University Press.

Davis, G. and C. Marquis. 2005. 'Prospects for Organization Theory in the Early Twenty-First Century: Institutional Fields and Mechanisms.' *Organization Science* 16: 332–343.

Davison, M. 2014. 'Red Tape Reduction: A Great First Step.' *Minerals Council of Australia*. Melbourne: Minerals Council of Australia.

Deegan, C., M. Rankin and J. Tobin. 2002. 'An Examination of the Corporate Social and Environmental Disclosures of BHP from 1983–1997: A Test of Legitimacy Theory.' *Accounting, Auditing & Accountability Journal* 15(3): 312–343.

Delmas, M. and M. W. Toffel. 2004. 'Stakeholders and Environmental Management Practices: An Institutional Framework.' *Business Strategy and the Environment* 13(4): 209–222.

DiMaggio, P. and W. Powell. 1991. 'The Iron Cage Revisited: Institutional Isomorphism and Collective Rationality in Organizational Fields.' *The New Institutionalism in Organizational Analysis*. W. Powell and P. DiMaggio, eds. Chicago: The University of Chicago Press, 63–83.

Dobbin, F. 2004. *The New Economic Sociology*. Princeton: Princeton University Press.

Eigen, P. 2007. 'Fighting Corruption in a Global Economy: Transparency Initiatives in the Oil and Gas Industry.' *Houston Journal of International Law* 29(2): 337–354.

Fligstein, N. 1991. 'The Structural Transformation of American Industry.' *The New Institutionalism in Organizational Analysis*. W. Powell and P. DiMaggio, eds. Chicago: The University of Chicago Press, 311–336.

Fligstein, N. and L. Dauter. 2007. 'The Sociology of Markets.' *Annual Review of Sociology* 33: 105–128.

Foucault, M., ed. 1998. *Ethics: Subjectivity and Truth*. New York: New Press.

Friedland, R. and R. Alford. 1991. 'Bringing Society Back In: Symbols, Practices, and Institutional Contradictions.' *The New Institutionalism in Organizational Analysis*. W. Powell and P. DiMaggio, eds. Chicago: The University of Chicago Press, 232–263.

Friedman, M. 1970. 'The Social Responsibility of Business Is to Increase Its Profits.' *The New York Times Magazine*.

Galaskiewicz, J. 1991. 'Making Corporate Actors Accountable: Institution-Building in Minneapolis-St Paul.' *The New Institutionalism in Organizational Analysis*. W. Powell and P. DiMaggio, eds. Chicago: The University of Chicago Press, 293–310.

Giddens, A. 1984. *The Constitution of Society: Outline of the Theory of Structuration*. Cambridge: Polity.

Global Reporting Initiative. 2010. 'Global Reporting Initiative.' www.globalreporting.org (accessed 2 March 2016).

——. 2014. 'Disclosure Database.' www.globalreporting.org (accessed 2 March 2016).

Global Reporting Initiative and United Nations Global Compact. 2010. 'Making the Connection: Using the GRI's G3 Guidelines for the UN Global Compact's Communication on Progress.' Amsterdam: Global Reporting Initiative, 30.

Granovetter, M. 1973. 'The Strength of Weak Ties.' *American Journal of Sociology* 78(6): 1360–1380.

Guthrie, J. and L. D. Parker. 1989. 'Corporate Social Reporting: A Rebuttal of Legitimacy Theory.' *Accounting and Business Research* 19(76): 343–352.

Haines, F. and A. Sutton. 2003. 'The Engineer's Dilemma: A Sociological Perspective on Juridification and Regulation.' *Crime, Law and Social Change* 39: 1–22.

Hall, P. and R. Taylor. 1996. 'Political Science and the Three New Institutionalisms.' *MPIFG Scientific Advisory Board Discussion Papers*. Klon: Max Planck Institut fur Gesellschaftsforschung, 32.

Hanlon, G. 2008. 'Rethinking Corporate Social Responsibility and the Role of the Firm: On Denial of Politics.' *The Oxford Handbook of Corporate Social Responsibility*. A. Crane, A. McWilliams, D. Matten, J. Moon and D. Siegel, eds. Oxford: Oxford University Press, 156–172.

Hannan, M. T. and J. Freeman. 1977. 'The Population Ecology of Organizations.' *American Journal of Sociology* 82(5): 929–964.

Hartman, L. 2008. 'Profitable Partnerships: How Multinationals Can Reduce Poverty While Driving Value.' *Annual Gourlay Lecture on Business Ethics*. Melbourne Business School.

Hartman, L., D. G. Arnold and R. E. Wokutch, eds. 2003. *Rising Above Sweatshops: Innovative Approaches to Global Labor Challenges*. Westport: Praeger.

Harvard University. 2011. 'Corporate Social Responsibility Initiative.' www.hks.harvard.edu/m-rcbg/CSRI/init_educ.html (accessed 2 March 2016).

Harvey, B. and S. Bice. 2014. 'Social Impact Assessment, Social Development Programmes and Social Licence to Operate: Tensions and Contradictions in Intent and Practice in the Extractive Sector.' *Impact Assessment and Project Appraisal* 32(4): 327–335.

Hedström, P. and R. Swedberg, eds. 1998. *Social Mechanisms: An Analytical Approach to Social Theory*. Cambridge: Cambridge University Press.

Hess, D. 2007. 'Social Reporting and New Governance Regulation: The Prospects of Achieving Corporate Accountability through Transparency.' *Business Ethics Quarterly* 17(3): 453.

Husted, B. 2005. 'Risk Management, Real Options and Corporate Social Responsibility.' *Journal of Business Ethics* 60: 175–183.

Hutchins, M. J., C. L. Walck, D. P. Sterk and G. A. Campbell. 2005. 'Corporate Social Responsibility: A Unifying Discourse for the Mining Industry?' *Greener Management International* (52): 17–30.

International Council on Mining and Metals. 2010a. 'Good Practice Guide: Indigenous Peoples and Mining.' www.icmm.com/document/1221 (accessed 1 April 2016).

——. 2010b. 'Sustainable Development Framework.' www.icmm.com/our-work/sustainable-development-framework (accessed 2 March 2016).

——. 2014. 'ICMM: Our History.' www.icmm.com/about-us/our-history (accessed 2 March 2016).

Jamali, D. 2007. 'A Stakeholder Approach to Corporate Social Responsibility: A Fresh Perspective into Theory and Practice.' *Journal of Business Ethics*: 1–19.

——. 2010. 'MNCS and International Accountability Standards through an Institutional Lens: Evidence of Symbolic Conformity or Decoupling.' *Journal of Business Ethics* 95(4): 617–640.

Jamali, D. and B. Neville. 2011. 'Convergence Versus Divergence of CSR in Developing Countries: An Embedded Multi-Layered Institutional Lens.' *Journal of Business Ethics* 102(4): 599–621.

Jepperson, R. 1991. 'Institutions, Institutional Effects, and Institutionalism.' *The New Institutionalism in Organizational Analysis*. W. Powell and P. DiMaggio, eds. Chicago: The University of Chicago Press, 143–163.

Jones, M. T. 1999. 'The Institutional Determinants of Corporate Social Responsibility.' *Journal of Business Ethics* 20: 163–179.

Joyce, S. and I. Thomson. 2000. 'Earning a Social License to Operate: Social Acceptability and Resource Development in Latin America.' *Canadian Mining and Metallurgical Bulletin* 93: 49–53.

Kang, N. and J. Moon. 2012. 'Institutional Complementarity between Corporate Governance and Corporate Social Responsibility: A Comparative Institutional Analysis of Three Capitalisms.' *Socio-Economic Review* 10(1): 85–108.

Kemp, D. 2010. 'Community Relations in the Global Mining Industry: Exploring the Internal Dimensions of Externally Oriented Work.' *Corporate Social Responsibility and Environmental Management* 17: 1–14.

Knight, J. 2001. 'Explaining the Rise of Neoliberalism: The Mechanisms of Institutional Change.' *The Rise of Neoliberalism and Institutional Analysis*. J. Campbell and O. Pederson, eds. Princeton: Princeton University Press, 27–50.

Kurtz, L. 2008. 'Socially Responsible Investment and Shareholder Activism.' *The Oxford Handbook of Corporate Social Responsibility*. A. Crane, A. McWilliams, D. Matten, J. Moon and D. Siegel, eds. Oxford: Oxford University Press, 249–280.

Kurucz, E., B. Colbert and D. Wheeler. 2008. 'The Business Case for Corporate Social Responsibility.' *The Oxford Handbook of Corporate Social Responsibility*. A. Crane, A. McWilliams, D. Matten, J. Moon and D. Siegel, eds. Oxford: Oxford University Press, 83–112.

Levy, D. and R. Kaplan. 2008. 'Corporate Social Responsibility and Theories of Global Governance: Strategic Contestation in Global Issues Arenas.' *The Oxford Handbook of Corporate Social Responsibility*. A. Crane, A. McWilliams, D. Matten, J. Moon and D. Siegel, eds. Oxford: Oxford University Press, 433–451.

Margolis, J. D. and J. P. Walsh. 2003. 'Misery Loves Companies: Rethinking Social Initiatives by Business.' *Administrative Science Quarterly* 48(2): 268–305.

Marquis, C., M. A. Glynn and G. Davis. 2007. 'Isomorphism and Corporate Social Action.' *Academy of Management Review* 32(3): 925–945.

Matten, D. and J. Moon. 2004. '"Implicit" and "Explicit" CSR: A Conceptual Framework for Understanding CSR in Europe.' *ICCSR Research Paper Series*. Nottingham University: ICCSR.

——. 2008. '"Implicit" and "Explicit" CSR: A Conceptual Framework for a Comparative Understanding of Corporate Social Responsibility.' *Academy of Management Review* 33(2): 404–424.

Meyer, J. W. and B. Rowan. 1977. 'Institutionalized Organizations: Formal Structure as Myth and Ceremony.' *The American Journal of Sociology* 83(2): 340–363.

Minerals Council of Australia. 2005. 'Enduring Value: The Australian Minerals Industry Framework for Sustainable Development.' Kingston: Minerals Council of Australia, 20.

Mining Minerals and Sustainable Development. 2002. Breaking New Ground: The Report of the MMSD: Earthscan.

Newell, P. 2008. 'CSR and the Limits of Capital.' *Development and Change* 39(6): 1063–1078.

Newmont Mining Corporation. 2014. 'Beyond the Mine 2013: Economic Development: Local Purchasing.' ed. Newmont Mining Corporation. Colorado: Newmont Mining Corporation.

Orlitzky, M., F. L. Schmidt and S. L. Rynes. 2003. 'Corporate Social and Financial Performance: A Meta-Analysis.' *Organization Studies* 24(3): 403–441.

Owen, J. R. and D. Kemp. 2013. 'Social Licence and Mining: A Critical Perspective.' *Resources Policy* 38(1): 29–35.

Özen, S. and F. Küskü. 2009. 'Corporate Environmental Citizenship Variation in Developing Countries: An Institutional Framework.' *Journal of Business Ethics* 89: 297–313.

Parsons, R. and K. Moffat. 2014. 'Constructing the Meaning of Social Licence.' *Social Epistemology* 28(3–4): 340–363.

Peru, Congreso de la Republica de. 2010. 'Ley Del Derecho a La Consulta Previa a Los Pueblos Indigenas U Originarios Reconocido En El Convenio No169 De La Organizacion

Internationale Del Trabajo ' In *1*, ed. Congreso de la Republica de Peru. Lima: Congreso de la Republica de Peru.

Porter, M. and M. Kramer. 2011. 'Creating Shared Value.' *Harvard Business Review* January.

Powell, W. and P. DiMaggio. 1991a. *The New Institutionalism in Organizational Analysis*. Chicago: The University of Chicago Press.

——. 1991b. 'Introduction.' *The New Institutionalism in Organizational Analysis*. W. Powell and P. DiMaggio, eds. Chicago: The University of Chicago Press, 1–38.

Power, M. 1997. *The Audit Society*. New York: Oxford University Press.

Prahalad, C. K. 2003. The Fortune at the Bottom of the Pyramid. Philadelphia: Wharton School Publishing.

Prno, J. 2013. 'An Analysis of Factors Leading to the Establishment of a Social Licence to Operate in the Mining Industry.' *Resources Policy* (4): 577.

Prno, J. and D. Scott Slocombe. 2012. 'Exploring the Origins of 'Social License to Operate' in the Mining Sector: Perspectives from Governance and Sustainability Theories.' *Resources Policy* 37: 346–357.

Reuf, M. and W. Richard Scott. 1998. 'A Multidimensional Model of Organizational Legitimacy: Hospital Survival in Changing Institutional Environments.' *Administrative Science Quarterly* 43: 877–904.

Rio Tinto Ltd. 2014. 'Sustainable Development Review 2013: Strong Partnerships for Success.' London.

Rokita, T. 2014. 'Red Tape Rollback: Annual Report/Second Edition.' In *Red Tape Rollback*. Lafayette, Indiana.

Sadler, D. 2009. 'Neo-Liberalising Corporate Social Responsibility: A Political Economy of Corporate Citizenship.' *Geoforum* 40(4): 613–622.

Salim, E. 2004a. 'Striking a Better Balance: Extractive Industries Review: Stakeholder Inputs: Converging Issues and Diverging Views on the World Bank Group's Involvement in the Extractive Industries (Volume 2).' *Consultation of the future role of the World Bank Group in the extractives industries*. Washington, DC: World Bank, 1–92.

——. 2004b. 'Striking a Better Balance: Extractive Industries Review: The World Bank Group and Extractive Industries (Volume 1).' *Consultation of the future role of the World Bank Group in the extractives industries*. Washington, DC: World Bank, 1–92.

Salzmann, O., A. Ionescu-Somers and U. Steger. 2005. 'The Business Case for Corporate Sustainability: Literature Review and Research Options.' *European Management Journal* 23(1): 27–36.

Schiavi, P. and F. Solomon. 2007. 'Voluntary Initiatives in the Mining Industry: Do They Work?' *Greener Management International* (53): 27–41.

Schmidt, V. A. 2008. 'Discursive Institutionalism: The Explanatory Power of Ideas and Discourse.' *Annual Review of Political Science* 11: 303–326.

Scott, W. R. 1991. 'Unpacking Institutional Arguments.' *The New Institutionalism in Organizational Analysis*. W. Powell and P. DiMaggio, eds. Chicago: The University of Chicago Press, 164–182.

——. 1994. 'Institutions and Organizations: Toward a Theoretical Synthesis.' *Institutional Environments and Organizations*. W. R. Scott and J. W. Meyer, eds. Thousand Oaks: Sage Publications, 55–78.

——. 1998. *Organizations: Rational, Natural, and Open Systems*. 4th edition. New Jersey: Simon & Schuster.

——. 2001. *Institutions and Organizations*. 2nd edition. Thousand Oaks, CA: Sage Publications.

Sethi, S. P. 2003. 'Proactive Corporate Responses: Voluntary Codes of Conduct.' *Setting Global Standards: Guidelines for Creating Codes of Conduct in Multinational Corporations*. Hoboken: John Wiley and Sons, Inc., 81–94.

Shapiro, D., B. I. Russell and L. F. Pitt. 2007. 'Strategic Heterogeneity in the Global Mining Industry.' *Transnational Corporations* 16(3): 1–35.

Smith, G. A. 2008. 'An Introduction to Corporate Social Responsibility in the Extractive Industries.' *Yale Human Rights and Development Law Journal* 11: 1–7.

Snider, J., R. P. Hill and D. Martin. 2003. 'Corporate Social Responsibility in the 21st Century: A View from the World's Most Successful Firms.' *Journal of Business Ethics* 48: 175–187.

Swedberg, R. 2003. 'Economic Versus Sociological Approaches to Organization Theory.' *The Oxford Handbook of Organization Theory*. H. Tsoukas and C. Knudsen eds. Oxford: Oxford University Press, 373–391.

Symonds, M. 2011. 'The Challenges of Tomorrow's World Today.' *The Independent*.

The Aspen Institute: Centre for Business Education. 2010. 'Aspen's Global 100: Beyond Grey Pinstripes 2009–2010.' New York: The Aspen Institute, 4.

The Economist. 2005. 'The Good Company.' 11.

Thelen, K. 1999. 'Historical Institutionalism in Comparative Politics.' *Annual Review of Political Science* 2: 369–404.

Thomas, G. 2004. 'Sociological Institutionalism and the Empirical Study of World Society.' *Observing International Relations: Niklas Luhmann and World Politics*. M. Albert and L. Hilkermeier, eds. London: Routledge, 72–85.

Treasury Board of Canada Secretariat. 2014. 'The 2012–13 Scorecard Report: Implementing the Red Tape Reduction Plan.' Ottawa: Treasury Board of Canada.

United Nations Global Compact and Accenture. 2010. 'A New Era of Sustainability: UN Global Compact-Accenture CEO Study 2010.' UNGC-Accenture, 60.

——. 2013. 'A New Era of Sustainability: Un Global Compact-Accenture Ceo Study 2013.' UNGC-Accenture, 60.

US Congress. 2010. 'Dodd-Frank Wall Street Reform and Consumer Protection Act.' In *HR 4173*.

Vogel, D. 2005. *The Market for Virtue: The Potential and Limits of Corporate Social Responsibility*. Washington, D.C.: Brookings Institution Press.

——. 2008. 'Private Global Business Regulation.' *Annual Review of Political Science* 11(1): 261–282.

Waddock, S. A. and S. B. Graves. 1997. 'The Corporate Social Performance-Financial Performance Link.' *Strategic Management Journal* 18(4): 303–319.

Weber, M. 1922/1946. 'The Social Psychology of the World Religions.' *From Max Weber: Essays in Sociology*. H. H. Gerth and C. Wright Mills, eds. New York: Oxford University Press, 267–301.

Wood, D. J., J. Logsdon, P. Lewellyn and K. Davenport. 2006. 'Global Business Citizenship: A Transformative Framework for Ethics and Sustainable Capitalism.' Armonk: M. E. Sharpe, Inc.

Yakovleva, N. 2005. 'Corporate Social Responsibility in the Mining Industries.' Aldershot: Ashgate.

Zucker, L. G. 1977. 'The Role of Institutionalization in Cultural Persistance.' *American Sociological Review* 42(5): 726–743.

3 The great adoption

Mining companies tell their CSR story

Corporate social responsibility is institutionalized in the global mining industry. But institutionalization does not necessarily result in permanent prioritization. Mining productivity sunk to global lows in 2014 (Deloitte 2014). Commodity prices are down. Iron ore prices dropped 47 percent in 2014 and are predicted to remain below US$60 for the foreseeable future. Newcastle thermal coal prices went down by 25 percent in the same year. Economist Ross Garnaut (2013) describes the shift as being one from 'salad days' to 'dog days'. Operating costs are up and the global industry saw a net loss in profits of 72 percent since 2013 (PricewaterhouseCoopers 2014). The phenomenal Asian growth referenced in Chapter 1 appears to be slowing. Although China's predicted deceleration to a rate of 7.6 percent in 2014 (The World Bank 2014) still well exceeds average global growth of 3.7 percent (The International Monetary Fund 2014), recent currency devaluations suggest a move away from commodity driven growth. Ore grades are also declining. Within this uncertain economic environment, the ICMM has prioritized attention to 'impact on community relationships' (International Council on Mining and Metals 2014) and leading global consultancy PricewaterhouseCoopers cautions miners about the risks to their social licence to operate related to growing resource nationalism and operating in frontier economies (PricewaterhouseCoopers 2014).

It is within this 'seismically shifting' context (Deloitte 2014) that mining companies' attitudes and actions toward global commitments and local communities are set. And we must ask, what are companies doing in response? It is critical to understand the current state of play of mining companies' non-financial responsibilities if we are to outline and explore the five principles to support responsible mining now and into the future. And given that these are principles which companies must adopt in order for them to be meaningful and successful, it is especially important to understand how companies see themselves. This is not to say that the perspectives and opinions of communities, governments, media and civil society are unimportant – indeed, Chapters 4 through 7 focus strongly on these groups. But examining companies' own perspectives provides a platform for understanding their perceptions of their values, roles and responsibilities.

Sustainability reports provide an excellent means of tapping this information, acting as the public face of mining companies' non-financial achievements and challenges. Chapter 2 introduced a lunchroom model of CSR. This analytical toolkit comprises four social mechanisms (shared language, peer pressure, group identity and discipline) and three levels of analysis (industry, corporate headquarter and operation site). In this chapter we apply the lunchroom model to expose the ways in which CSR[1] has been institutionalized at the industry (organizational field) level. In the discussion that follows, we apply those social mechanisms to 50 multinational mining company sustainability reports, representing five companies across a decade (2003 to 2013).[2] Companies analysed include BHP Billiton, Rio Tinto, Newmont Gold, MMG (formerly MinMetals, formerly Oxiana) and Xstrata. The content and discourse analysis of these reports is bolstered by 51 interviews completed in Australia between 2009 and 2010 with multinational mining company executives, senior managers, community relations staff, contractors and community representatives living alongside mining operations.[3]

Putting sustainability reports in context: recent experiences of CSR

Beginning early in the twenty-first century, mining companies progressively adopted annual sustainability reporting practices. This movement reflects a broad business trend in which sustainability reports annually recount firms' achievements and shortcomings related to social, environmental, labour and human rights behaviours. In 2013, 93 percent of the world's largest 250 firms produced a sustainability report (KPMG International 2013). And sustainability reporting is no longer limited to the largest companies. Or to Western companies. Reporting by companies based in the Asia-Pacific increased to 71 percent in 2013, up from 49 percent in 2011 (KPMG International 2013).

Recent writings of industry leaders and senior corporate representatives provide important context to the rapidity and extent to which CSR has diffused across the global mining industry in recent years. Bruce Harvey, Rio Tinto's former Global Practice Leader: Communities for almost 30 years until 2014, suggests that although Rio Tinto had a longstanding and developing CSR practice, it was only in the mid-1990s that the company underwent internal cultural change toward its social responsibilities, especially to indigenous Australians (Harvey 2006). In a speech to the Minerals Council of Australia (MCA), Harvey attributed this noteworthy shift to 'top-down' influences, including the MCA's *Enduring Value Framework* for sustainable development. The timing of this shift in corporate attitude was also very likely influenced by Australia's 1992 Mabo case decision concerning native title which overturned longstanding claims of *terra nullius* to acknowledge traditional landowners' rights (Harvey and Nish 2005). Carolyn[4] (2009), a community relations senior manager I interviewed, cited other reasons altogether, attributing a growing interest in CSR to shifting economic circumstances and stakeholder desires. She suggested this change was evident in fairly recent alterations in her company's budget allocations:

> Historically, if you look 10 to 15 years ago [1999–1994], 'spare' money in the budget was put towards professional development. Professional development spending has generally decreased and the focus has shifted towards putting this money towards community, so the money's been redirected. … The increased investment in the community also has to do with broader economic cycles and community expectations.

Diane (2009), a senior manager of sustainable development for global operations, also pinpointed the diffusion of corporate social responsibility across the mining industry as peaking within the past decade or so. She linked increased interest in CSR to rising public awareness of environmental issues. She noted that this was evidenced in her home country of Australia through the political elevation of climate change as a mainstream issue.[5] While she believes that companies' concern for community issues lags behind their interests in environmental matters, she suggests that social concerns will become better integrated wherever community actions are acknowledged as a threat to business operations:

> Environment has positioned itself about 10 or 15 years ago as central. But community is going to increasingly be considered as we go into tougher and tougher environments. If you don't get the community on side, then you're facing community opposition that can stop production.

Catherine (2009), a senior advisor for community relations, echoed both a similar timeline for the recent surge in CSR and reasons analogous to Diane's for its rapid diffusion through the industry, saying:

> Even if you look at environment 10 or 15 years ago, no one gave a toss. But now our values are changing around the environment. And I think values around community are shifting as well.

She (Catherine 2009) went on to suggest that a growing concern about social issues is linked not so much to the potential harm which can be done by a community in opposition to an operation, but to apprehension about the effects on productivity which may occur where a local community does not have the capacity to support the mining workforce:

> I find it really hard to nut down [what the industry's current CSR priority is] because I think it can really take a few different bows. But if you look at [a certain mining operation], they had huge issues where they couldn't even get the workforce to work because of child care issues. So, how do you help the community to build capacity to become more productive?

Interviewees' assertions that interest in CSR from the mining industry has grown in recent years is supported by researchers' investigations into local and global trends. Craig Deegan and colleagues (2002) identified a recent surge in attention to socially responsible behaviours through examining BHP Billiton's annual reporting. They linked this upswell partly to increased media coverage and remote communities' improved abilities to garner public attention. Authors like Hevina Dashwood (2004) and Ralph Hamann (2004) identify a similar trend on a global scale, attributing recent changes in mining companies' actions toward community and environmental issues to mounting pressure from NGOs, government and community activist groups, as well as from within the mining industry itself. Dashwood's (2004 and 2007) exploration of the Global Mining Initiative, and consequent ICMM, for instance, suggests peer pressures through which companies have pushed one another to adopt socially responsible practices for the greater benefit of the industry.

The recent diffusion of industry-wide corporate social responsibility initiatives reveals one of the major contradictions characterizing CSR in mining. A 'benefits for all' mantra has been an important catalyst for adoption of these initiatives. Yet many historically competitive companies find the cooperative relationships necessary to advance these agendas alien. In Australia, for example, staunchly competitive relationships between companies arise not only from obvious capitalist motives, but from the common circumstance of several mining companies operating next door to one another within a mineral-rich region (Franks, Brereton et al. 2010).[6] In these situations, companies not only compete for limited resources available within adjacent places, they also compound and influence the social and environmental impacts of their neighbours' operations. Adoption of a more cooperative approach to address mining's cumulative impacts is needed (Franks, Fidler et al. 2009). But it is difficult.

Ongoing reticence to collaborate may also be associated with a history wherein community programmes were instituted for the perceived reputational benefits to be gained (Centre for Social Responsibility in Mining 2007). I have seen this style of programmes first-hand in communities where I have undertaken fieldwork. On one fieldtrip in Australia, my morning jog was punctuated by prominently placed placards dating from the late 1990s reminding me that the footpath along which I ran, the lookout where I appreciated the view and the playground where I paused to stretch were 'Brought to [me]' by one particular company. This is a scene I have seen repeated in many places around the world. While somewhat anachronistic today, these little signs emblazoned with the company's logo and detailing the date and contributions of the company to community infrastructure serve as a longstanding reminder of the depth to which the company is entrenched within the community. But as impacts of mining accumulate within a geographic area and companies begin to work more collaboratively to address their regional impacts, the types of company contributions are changing and it is becoming difficult for any one company to claim this type of credit for community benefits. As Susan (2009), a general manager, sustainable development, explained, where multiple companies operate in a particular area and cooperate on community initiatives, individual companies lose the ability to determine or claim their community contributions:

> [It's challenging] if you're working in an area where you've got lots of companies and lots of organizations and government and everyone's pulling their weight to address an issue, and you can't really determine whether your work is what's caused the change.

Susan went on to explain that her company is not overly concerned with garnering accolades if the ultimate result of cooperative community programmes is an outcome like malaria eradication. She also acknowledged that it can be difficult for her and her team of employees to earn board-level support for projects with results that are difficult to credit specifically to their own company. The situation is complex and changing. Susan also noted that her company's increasing willingness to collaborate on projects with other companies or to undertake more stakeholder engagement is based on the creeping realisation that the types of reputational gains associated with more traditional CSR practices are perhaps not as valuable as historically perceived:

> Because we think, in the longer term, you have to do … things for people to actually value your having been there. It's nice to see all the money we've put into things through the project, but the community won't identify it with us after we've gone if we haven't built the relationship through the project.

Mining companies' approaches to their non-financial responsibilities are evolving. And sustainability reports provide an interesting and often untapped record of this development.

Sustainability reports and the public face of the institution

Sustainability reporting has now been adopted to such an extent within the industry that it is safe to infer that non-reporting may negatively influence a firm's reputation (Ernst and Young and Boston College, Carroll School of Management Center for Corporate Citizenship 2014). Although sustainability reporting remains voluntary in many countries, there is an emerging sense that miners now have little choice but to produce them to demonstrate their CSR commitments publicly. Researchers Petrina Schiavi and Fiona Solomon (2007) connect companies' concerns with social responsibility to voluntary initiatives, like the Global Reporting Initiative (GRI), proliferating since the early 1990s. They suggest that companies' interests were initially piqued because of the frameworks' potential reputational benefits.

Today, sustainability reports are front and centre in mining companies' communication. Each of the companies studied in this chapter links directly to their current and past sustainability reports from their online homepages through portals variously named 'society', 'our commitment', 'sustainability and community' or 'sustainability'. The reports are disseminated annually, aimed at a wide audience, including shareholders, employees, community members, investment analysts, contractors, suppliers, NGOs and government. Much like annual financial reports, sustainability reports follow a generally accepted formula for disclosure and usually include the following: statements from the chief executive officer, board chair or senior executives about how their company approaches corporate social responsibility; outlines of any voluntary or involuntary commitments which the company has made relative to CSR; statistics and case studies about environmental, social, economic, employment and governance issues; and performance concerning those issues and future performance targets.

Ideally, sustainability reports offer transparent communication which discloses where the company has faced challenges, failed in its commitments or where it struggles to implement policies on-ground. Research suggests, however, that this type of disclosure is where sustainability reports generally come up short, with content focusing instead on positive outcomes or noting the impacts of a corporation without indicating how it addresses those impacts in practice (KPMG International and SustainAbility 2008). Consequently, although sustainability reports ostensibly aim to provide stakeholders with transparent information about a company's social, environmental, economic and governance impacts, report readers believe that companies are unlikely to report failures (KPMG International and SustainAbility 2008). Even among the mining company representatives I interviewed (Walt 2009), scepticism about the alignment of mining companies' public communications and actual practices was prevalent:

> One of the major take-aways [of the Newmont 'Global community relationships review'] was the fact of this distance between corporate speak in respect to sustainability and community relations, and what actually happens on the ground. ... You can go to any website and have a look at the sustainability drop-down menus and you think, 'What a fantastic company!' But you get out on the ground and there's just such a disparity.[7]

Sustainability reports must be approached somewhat cautiously, and with their specific goals in mind. Ultimately, they are corporate publications, vetted by internal experts, marketing and communications teams, and legal advisors prior to publication.[8] As such, they provide a window onto how these firms wish to be perceived, relative to their social, environmental, economic and governance responsibilities. While many sustainability reports are independently 'assured' by sustainability auditors, no formalized auditing procedure yet exists to support more objective assessments and comparative analyses (Frost and Martinov-Benny 2010). The legitimacy of external assurance of the reports becomes particularly important when we consider whether stakeholders can trust sustainability report data.

Greenwashing and marketability

Accusations of 'greenwashing' tailgate mining companies' sustainability reporting. One recent study found fair grounding for these concerns, noting that the GRI's organizational-level focus may result in data aggregation that paints a brighter picture than its individual parts. The research also found shortcomings in assurance standards, especially where they allow high degrees of management control over what issues are investigated in the assurance process, and where engagement with stakeholders' concerns is limited (Hall 2014). We will return to the issue of assurance in greater detail in a moment. The 'specious gloss' applied to sustainability reports' presentation also spurs allegations of greenwashing (Hall et al. 2015, 253). Slick presentation raises suspicion that important issues are disguised or that only the positive sides of very complex stories are told. In some instances, this is the case (Global Reporting Initiative, The University of Hong Kong et al. 2008).

Greenwashing is most certainly a concern when it comes to sustainability reporting. But glossiness and meaningful disclosure are not mutually exclusive. It is worth retiring cynicism for a moment to consider the various other reasons why companies may produce a well-designed report. A sustainability report is a public representation of the corporation and reflects corporate style guides and expectations for public documents (KPMG International and SustainAbility 2008). The production of sustainability reports requires a great deal of time and effort for compilation and publication and it is a fair assumption that those who have expended energy gathering data and composing the report wish for it to attract readership.

It is also important to remember that many sustainability reports, especially those of the mining companies studied here, are disseminated to extremely diverse audiences with a broad range of educational backgrounds, first languages and literacy levels, and with different reasons for interest in the companies. Financial analysts increasingly look to these reports to alert them to material corporate risks (Frost and Martinov-Benny 2010). Local community members are more likely to seek out information about how the company is affecting their region and any steps taken to mitigate impacts (KPMG International and SustainAbility 2008; KPMG International and SustainAbility 2010). Other companies take interest in the activities of their competitors. Each of these factors influence report design.

Many companies do view the sustainability reports as marketing documents and want to present an engaging image. Companies are also aware that few readers will read the reports cover-to-cover, and the use of photographs and charts facilitates the type of skim reading commonly used for reports which are often lengthy and usually dense (KPMG International and SustainAbility 2008).[9] Photographs and charts may also make a story more accessible to those stakeholders for whom English is not a first language or who may have limited literacy. Sustainability reports' appearance and style are not as important as the information disclosed but these factors comprise an important aspect of CSR discourse. Even down to companies' choice of recycled paper, vegetable-based inks or provision of an online-only report, the format and appearance of the reports suggests something about companies' understanding of, approaches and commitment to CSR (KPMG International and SustainAbility 2008).

External assurance

It is perhaps too easy to write off reports as greenwash or as cynical attempts to win over stakeholders or to present a positive but unrealistic public image (KPMG International and SustainAbility 2010). Instead, it seems more useful to consider how the adoption of international CSR reporting frameworks in recent years has helped improve the veracity and balance of content through the uptake of external report 'assurance' – an independent auditing of CSR data (KPMG International and SustainAbility 2008). By 2004, four of the five companies studied adopted external assurance as part of their sustainability report writing process. By 2013, all companies had their reports externally assured. Since 2010, ICMM members have ascribed to the ICMM Assurance Principle which requires statements on compliance with the ICMM *Sustainable Development Framework*, production of a GRI G.3 A+ report (now G.4) and third-party report assurance following GRI guidelines (Williams, Gill et al. 2007). Other voluntary standards such as the AA1000 Assurance Standard (AccountAbility 2008) also inform report assurance.

Sustainability report assurance is usually undertaken by consulting firms which specialize in CSR and which have strong familiarity with the GRI and AA1000 Standards (Frost and Martinov-Benny 2010). The global 'Big four' accounting firms, however, are increasingly moving into the sustainability report assurance market, lending their auditing expertise and the weight of their imprimaturs. Assurers frequently work in small teams in which individual members bring expertise in environmental sciences, social sciences and, occasionally, economics (Frost and Martinov-Benny 2010). Assurance statements provide report readers with a third-party assessment of the objectivity and balance of the reports and also point out areas in which companies may improve their CSR disclosures in future years.

Research shows that assurance statements instil a degree of trust in report content and address report users' desires to see external verification of information and consequent transparency (KPMG International and SustainAbility 2008). Although the majority of reports studied here – especially those produced after 2010 – were independently verified, it is still important to keep the corporate origins of each sustainability report in mind. Ultimately, sustainability reports offer a wealth of information about the social drivers shaping CSR in the mining industry, and the consequent implications of companies' public attitudes toward and actions against corporate social responsibility.

The following sections detail the results of a content and discourse analysis of five multinational mining companies' sustainability reports published in the decade 2004–2014.[10] In total 50 reports covering over 1 million words and approximately 4,319 pages were analysed. We searched for both 'manifest' (i.e. those readily observable because they are communicated using synonymous words) and 'latent' themes (i.e. those identified through interpretation) (Neuendorf, 2002). For example, while reports may not have stated explicitly that a particular case study related to issues of gender inequality or employee retention, this could be inferred through discussion of programmes to boost women's employment or a chart containing turnover rates, respectively. Details of the method and analysis appear in the appendices.

Defining what matters: a shared language for corporate social responsibility

Sustainability report content tells us much about the creation of shared language to discuss the mining sector's social, environmental and economic impacts and perceptions of responsibility to address these. The way in which particular topics are presented also reveals much about how the industry understands and is equipped to address particular concerns. For example, tensions are especially visible where mining companies attempt to address largely intangible social issues from within an industry that is historically and pragmatically reliant on measurement. This often results in quantifiable data shaping both the concept and practice of companies' responsibilities, effectively creating a self-reinforcing feedback loop: the things companies can measure clearly and reliably are those which are prioritized and which move by centripetal force toward the centre of companies' responsibility agendas. The adoption of shared language and changes in the degree to which particular issues are discussed in the reports across the studied years reveals isomorphic pressures introduced in the lunchroom model in Chapter 2 (refer back to Figure 2.1).

The earliest sustainability reports included in the content analysis discuss corporate social responsibility primarily in terms of environmental issues (47.0% of total CSR disclosures in 2003–2004) and employment matters (46.5% of total CSR disclosures in 2003–2004), especially relative to occupational health and safety (OHS) (47.9% of employment matters content in 2003–2004). This is very similar to findings produced by scholars James Guthrie and Lee Parker (1989) in a study focused on BHP Billiton which predates my own by about a decade. Is the lack of change in reporting content across this period coincidental? Or are these peer pressure and group identity processes writ large?

In 2003–2004, three of the five reports produced for that year went by titles other than 'Sustainability Report', such as 'Corporate Report' or 'Health, Safety, Environment and Community Report'. By 2005–2006, four out of the five companies included in the sample adopted the title 'Sustainability Report', with the remaining company calling its report a 'Sustainable Development Review'. This titular change is not trivial. It represents a coalescing discourse around corporate social responsibility, at least for these five firms. The timing of this shift in report titles to 'Sustainability Report' was likely influenced by the 2006 release of the third and most comprehensive version of the GRI, the G3 Sustainability Reporting Guidelines (GRI G3). This version of the GRI framework is either wholly adopted or at least referenced as a touch-point for guiding report content in all reports studied from 2006, with the earlier reports studied having at least referred to previous versions of the GRI.

Sustainability report content also began to change at the same time that a more unified discourse of sustainable development appeared across the mining industry. For instance, the companies analysed began to shift their corporate social responsibility disclosures from primarily environmental and OHS concerns to more social issues. Several companies explicitly linked this shift in report content to their use of more advanced external sustainability indicators such as the GRI G3, which provided them with more detailed guidance about the specific social issues with which a responsible company might be concerned. Several companies also noted the importance of other internal and external factors, such as a growing awareness of stakeholder expectations, increasing amounts of partnership work with NGOs, commitment to voluntary frameworks or industry and associated guidelines, and changes in board governance structures to accommodate sustainable development committees. In BHP Billiton's 2005 'Sustainability Report', for example, the company credited its incorporation of more social issues to a desire to align better with the GRI. The change is also reflected in its then-newly improved internal understanding of corporate social responsibility and acknowledgement of stakeholder concerns. Consequently, BHP Billiton explicitly earmarked 2005 as a vital turning point in its approach to public disclosure on non-economic issues. The report states:

> Our 2005 Sustainability Report represents the next step change in this evolution. The change has been a result of the increasing breadth of our reporting, reflecting the maturing of BHP Billiton's approach to sustainable development through the improved integration of social, environmental, ethical and economic factors into all that we do.

As with BHP Billiton, other studied companies' changing perceptions about the issues and concerns that relate to non-financial responsibilities appear to have been influenced by reporting guidelines and stakeholder input. These perceptions were reflected not only in sentiment, but in a marked change in companies' report content across the sample. In the 2003–2004 sustainability reports studied, social issues were given short shrift, making up 16.8 percent of corporate social responsibility issues disclosure content, compared to approximately 41 percent of total disclosures content for both environment and employment related disclosures. By 2006–2007, the proportion of reporting on social issues had increased over 1½ times, with social issues comprising 26.9 percent of CSR related disclosures (see Figure 3.1). While the proportion of reporting on social issues has since levelled off (in 2013 social issues comprised 23 percent of all CSR disclosures), the number of social disclosures has increased across the board. In line with this, environmental and employment issues continued to make up the bulk of corporate social responsibility disclosures, comprising 40 percent and 37 percent of total CSR disclosures, respectively in 2013–2014.

The numbers pointing to inclusion of more social issues suggest an important shift in thinking about the types of concerns which affect mining companies' responsibilities to society. But it is difficult to confirm by numbers alone, especially because growth in social issues reporting occurred alongside broader inclusion of a wider range of sustainability issues in all categories over the years studied. While the proportion of report content dedicated to social issues did increase and level off during the period, the reports' CSR disclosures did remain primarily focused on environmental and employment issues, although these issues' proportion of total coverage did decline somewhat (41.4% to 37.0% and 41.8% to 36.1%, respectively).

The content analysis shows that 40 percent of total CSR disclosures from all years focused on environment issues, followed closely by employment issues at 37 percent. Conversely, in all studied years combined, social issues made up just under one-quarter (23%) of CSR disclosures, approximately 17 percent less coverage than environmental or employment issues during the period. The

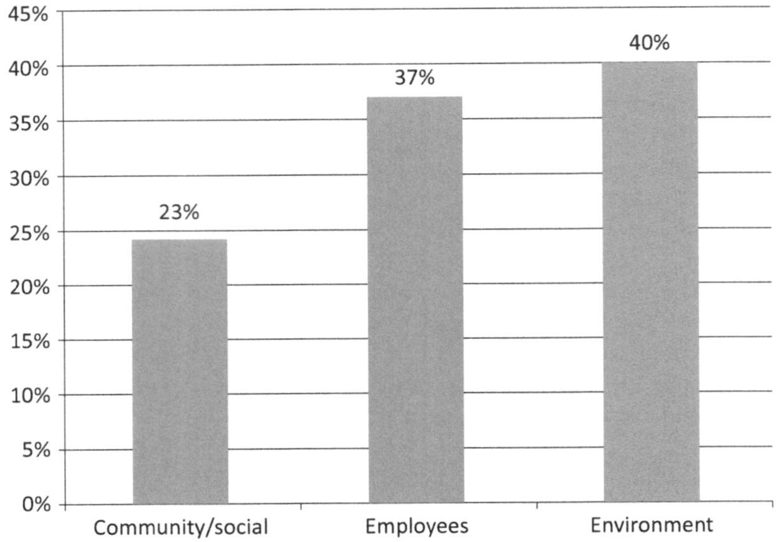

Figure 3.1 Comparison of total social, employment and environmental CSR disclosures: 2013–2014

disparity between the total amount of social issues content and total environmental and employment issues content can be explained in part by the very low coverage of social concerns in earlier years, which skews the proportion of total content. Even when this is taken into consideration, however, social issues have never achieved the same levels of coverage within companies' sustainability reports as other disclosures.

While reporting on social issues has increased, the reasons behind this shift remain unclear. Previous studies have investigated whether changes in report content can be linked directly to external events, such as major environmental incidents or media coverage (Guthrie and Parker 1989; Deegan et al. 2002; Guthrie and Yongvanich 2005). But these studies have been inconclusive about a linkage between the two. Guthrie and Parker's early study (1989) could not substantiate linkages and was entitled 'A rebuttal of legitimacy theory' for this reason. Craig Deegan and colleagues (2002) found a stronger relationship, but their measure of social issues is somewhat problematic in that media coverage is not necessarily an appropriate proxy for identifying community concerns. Perhaps a more interesting question is: why do companies decide to focus on particular types of issues over others (e.g. environmental vs. social) within reports? How do carefully selected discourses institutionalize a particular form of corporate social responsibility among major mining companies?

Professor of corporate responsibility Michael Blowfield (2007) argues that the lack of focus on social issues within sustainability reports occurs because of a reliance on somewhat vague case study methods to communicate social impact information. Case studies are limited in their ability to: pinpoint specific social impacts; connect them in a direct and meaningful way to key stakeholders; or demonstrate the effectiveness of sustainable development programmes. These are all arguably important components of a shared language for CSR.

The rather vague qualitative discourse around social issues in the sustainability reports studied here reveals a prizing of quantitative or quantifiable data over qualitative information or 'storytelling'. In their 2006 sustainability report, BHP Billiton, for example, included 33 case studies; 23 of which related to social or community issues. The same report included 132 quantitative charts and graphs, of which less than 10 percent represented social or community issues. Of the ten quantitative charts and graphs covering social issues, two concerned community complaints, one quantified the types of issues discussed during company-stakeholder meetings, one recorded the number of arrests over time in an area in which the company operates, and the remainder recorded monetary contributions related to communities.

Patterns like those described above occurred across all reports studied (see Figure 3.2). Environmental issues were twice as likely to be presented in a quantitative manner than social issues (82.2% compared to 40.2%). For example, Newmont's (2005) 'Environment' section of its report is jam-packed with graphs detailing energy consumption, CO_2 and other types of air emissions, water use and mineral waste production figures. By contrast, the 'Community relations and

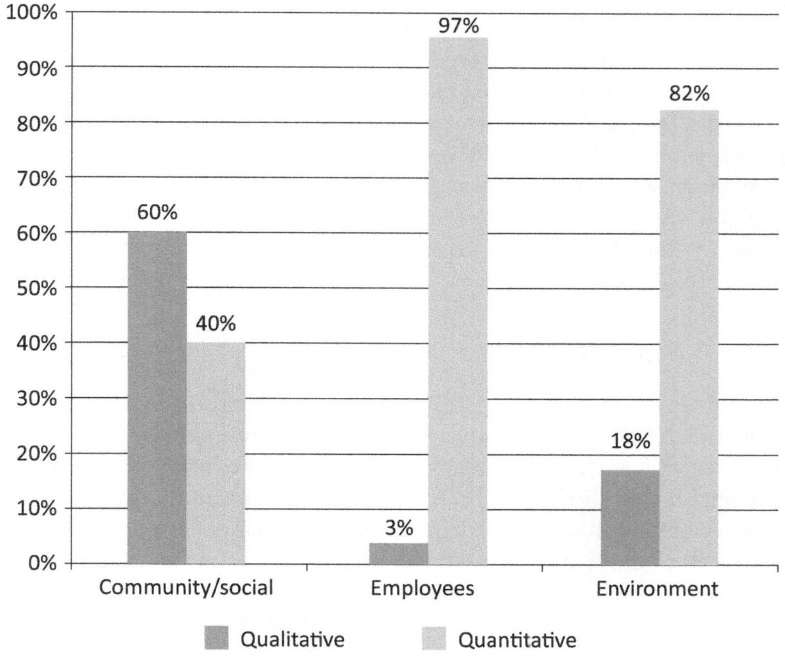

Figure 3.2 Environmental, social and employment issues disclosures: all years combined – qualitative vs. quantitative mentions

development' section is filled with qualitative data, largely in the form of case studies and contains only three graphs, each of which refers to money invested in local communities. This pattern of quantifying environmental, economic and labour data while qualifying social data was consistent across all reports. Employment issues were even more strongly quantitative than environmental. About 97 percent of employment issues were presented quantitatively (see Figure 3.2), employing charts and graphs showing number and diversity of employees, turnover rates and total injury frequency rates. In general, social issues displayed a pattern inverse to environmental or employment issues reporting, with the amount of qualitatively communicated information (59.8%) greater than the amount of quantitatively communicated information (40.2%). The GRI and CSR Asia made a similar finding in their 2008 study of sustainability reports across several industries, noting that companies use less quantitative indicators for reporting community impact information as compared with other CSR disclosures.

The prevalence of quantitative over qualitative discourse in the mining industry is not surprising. It links to the industry's roots in the scientific languages of geology and engineering. Mining industry representatives I interviewed indicated that the esteem for quantitative information over qualitative represents a values-system intrinsic to the sector. Even outside of charts and graphs, the studied reports favour a pseudo-scientific discourse within the report text, relying heavily upon notions of 'measurement', 'targets' and 'indicators', instruments which by their nature are more amenable to a quantifiable language associated with science and resources.

The finding that sustainability report disclosures appear to value environmental and economic concerns over social matters may reflect common industry language and concerns. But this finding does not align with companies' qualitative assertions about their commitments to social issues. The content analysis suggests instead a loose coupling between companies' publicly stated concerns about social issues and impacts and the amount of information they produce about those issues, compared to other concerns. In the section below, we turn our attention to how this discrepancy might be explained.

Where is the social in CSR?

The amount of sustainability report disclosures dedicated to social issues reveals an irony critical to the mining industry. Those issues most reported on in relation to corporate *social* responsibility are not so much social by nature, but environmental or economic. If CSR disclosures began shifting to include more equal coverage of social issues around 2006, why has reporting on social issues not reached equal content levels to those of environmental or employment matters today?

An initial interpretation of the findings might be that in the cases of the mining companies studied, social issues have not been prioritized to the same degree as environmental or employment concerns. Yet the spread of a shared language of responsibility across the industry, evidenced both by sustainability reports studied and interviewees' feedback on their companies' approaches to CSR, suggests this is not the case. In their 2006–2007 'Sustainable development review', Rio Tinto Ltd (2007), for example, states that social issues are just as integral to its CSR, based on what it understands to be the business case for attending to social responsibilities and to the reputational consequences related to how it deals with social, environmental and economic considerations:

> Ethical integrity and social responsibility are critical to the way we conduct our business. We aim to build enduring relationships with our stakeholders that are characterised by mutual respect, active partnership and long term commitment. In the long run, the trust that is engendered by solidly based relationships will reinforce Rio Tinto's ability to gain preferential access to resources.

In this instance, both business case and legitimacy concerns influence CSR commitments. Similarly, in their 2006–2007 'Full sustainability report' BHP Billiton (2007) includes, 'effectively addressing heightened political and societal expectations related to the environmental and social aspects of our business', as a vital component in earning and maintaining their social licence to operate. Without attention to social issues, as well as environmental and economic matters, BHP Billiton states that it will not be assured of the access to resources it requires to be a successful business. Taken together, the sustainability reports suggest that these companies deem social matters to be equally important to their environmental, economic and governance concerns, even if only because they have the potential to affect business viability, as in the BHP Billiton example. Yet the content analysis shows that social issues do not receive equal space within the reports and that social matters tend to be presented in language and formats different from those used to discuss environmental and employment issues.

The GRI and CSR Asia's (2008) recent study of companies' community impacts disclosures in sustainability reports argues that most major corporations (across several industries) generally lack awareness of their social impacts. Consequently, they do not report heavily on these issues. Compared to other industries, the mining sector appears more advanced in its reporting on social issues, although social content still lags attention to other sustainability concerns. Figure 3.3 shows this pattern reflected among the mining companies studied. The social issues most reported by companies across all years relate to community health, community development programmes, social impacts and the delivery of value to communities. Of these, the 15 percent that focused on community health tended to be quantified, in line with interviewees' opinions that health concerns are those most easily quantifiable. Community development programmes, however, were discussed slightly more than health issues (16%) and were presented mostly qualitatively and through case studies. Other main social issues reported on included 'social/ community impacts' (14%) and 'value to communities' (13%), both of which tended to be presented quantitatively, the latter in monetary figures and the former in terms of number of programmes implemented or community members attending those programmes.

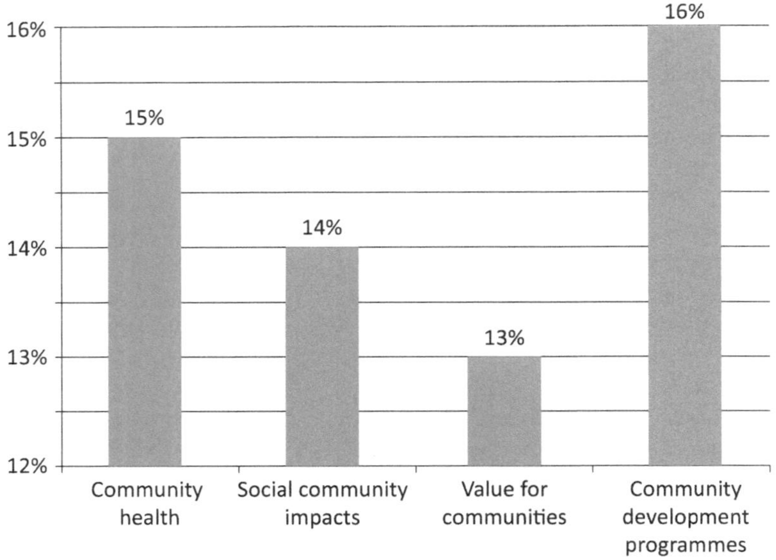

Figure 3.3 Community/social disclosures: all years combined

Issues such as gendered concerns – which are widely studied in academic, mining-related development literature (see, for example, Emberson-Bain 1994; Kunanayagam 2003; Lahiri-Dutt and Macintyre 2006) but notoriously difficult to quantify (Rio Tinto 2009) – do not appear in the sustainability reports as often. Several mining company representatives interviewed suggested that this was because these issues are deemed difficult to measure or are seen as 'soft' and therefore marginal issues for their companies. When separated out from social issues, gender concerns comprise only three percent of total CSR disclosures across all years combined (see Figure 3.4) and receive the least attention among other social issues presented in the sustainability reports (1% of total social issues disclosures, all years combined).

Figure 3.4 also demonstrates the various employment and environmental issues covered in the reports and the extent to which each issue appears. The figure demonstrates which CSR disclosures are prioritized in terms of report content and helps us to understand better which issues the studied companies believe matter most when it comes to making CSR disclosures. For instance, Figure 3.4 clearly reveals the difference in the level of reporting on social compared to other CSR disclosures, providing a further breakdown of more detailed issues, such as 'OH&S', 'value for communities' and 'waste management', which comprise each broader category. For each major issue set, 'environment', 'social' and 'employment', the percentages given represent the proportion of total mentions of that type of issue across all years of the study (2003 to 2014). Among environmental issues, for example, 'greenhouse gas/carbon emissions and climate change' comprised the most frequently mentioned environmental concern across all companies and years (17%), followed closely by reporting on companies' water management (15%).

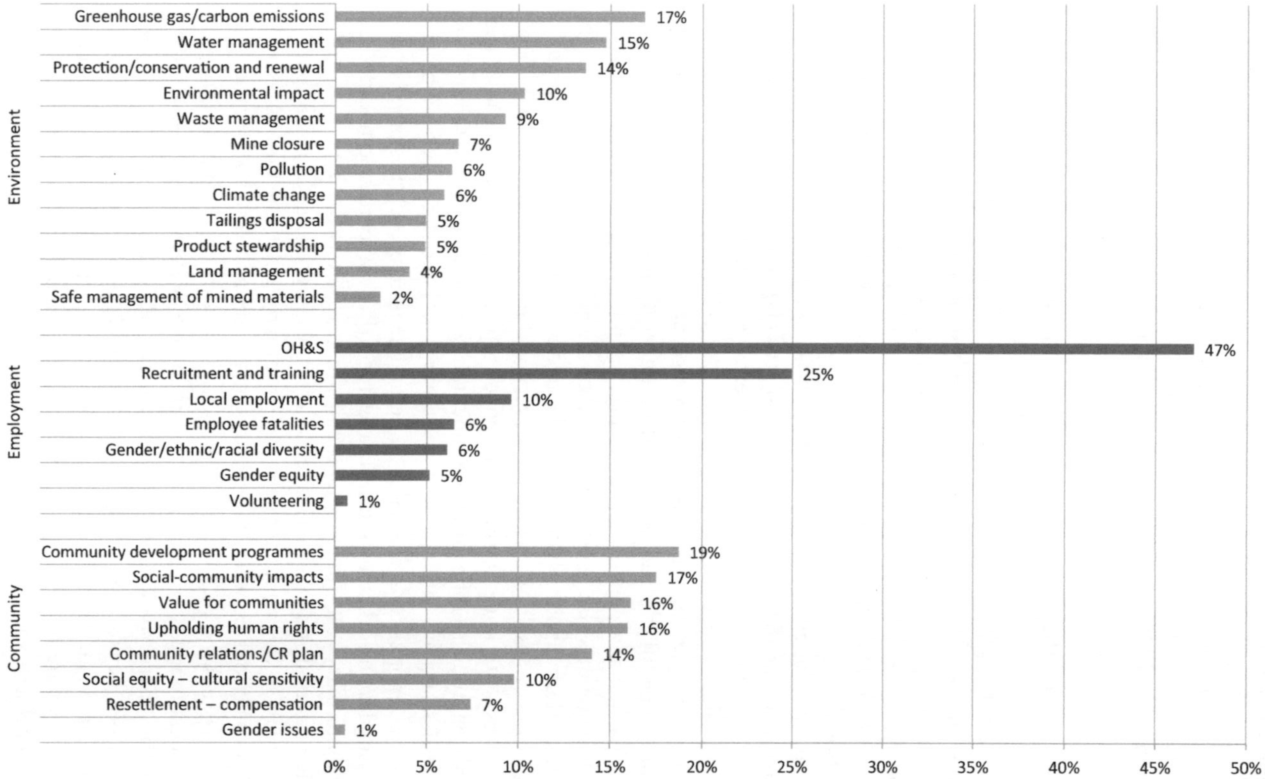

Figure 3.4 Breakdown of employment, community and environmental issues reported: all years combined

This environmental information is largely presented in charts or tables with related, descriptive text. For instance, greenhouse gas/carbon emissions and climate change usually merit discussion of each company's beliefs about the impacts of emissions and their approach to abatement. In 2005, for example, when the certainty of climate change remained widely debated, Rio Tinto Ltd stated:

> Rio Tinto believes that emissions of greenhouse gas resulting from human activities are contributing to climate change. … Our climate change position was revised in 2005 to take into account our improved understanding of this topic.

OH&S was by far the most discussed employment issue (47%), followed by information on recruitment and training (25%), the discussion of which frequently referred to staff turnover statistics and participation rates in apprenticeship and trainee programmes. These discussions are largely quantitative, are usually paired with graphs or tables and are primarily descriptive. MMG's 2006 (Oxiana 2006) (39) report typifies this where it states: 'A total of 123 people left employment at the Sepon Operation during 2006. Staff turnover was 26% for expatriates and 10% for Lao nationals.'

As the majority of environmental and employment issues were presented in a quantitative manner (82.2% and 97.1%, respectively), the level of attention to particular environmental or employment concerns over others cannot be attributed solely to the ability to quantify them. Instead, and different to social issues, employment and environmental issues coverage appears to be shaped less by which data is quantifiable and more by the degree of importance each company places on a particular aspect of environmental or employment concerns.

Many of the reports also state that issues that are given greater attention in the reports reflect those about which their stakeholders' have expressed concerns or which the company itself has prioritized in recent times. BHP Billiton (2007), for example, states: 'We have attempted to cover topics and [GRI] indicators that reflect BHP Billiton's significant environmental, economic and social impacts or that would substantively influence the assessments and decisions of stakeholders.' In most of the reports studied, such a 'material issues' approach is shaped by the GRI reporting framework, which encourages reporting companies to undertake a five-step 'materiality test' prior to report drafting, to focus the report on those issues which internal and external stakeholders care about most (Global Reporting Initiative 2006). In this way, we might expect initiatives like the GRI to result in improved connections between corporate headquarter policies and operation site activities. Yet studies like Matthew Genasci and Sarah Pray's (2008) on socio-economic accountability in the extractive industries suggest otherwise. Instead, they argue that extractives industry companies pay limited attention to particular social issues not because of a lack of awareness of materiality, but due to the 'distraction paradigm.' Here, mining and extractive companies focus attention on 'lower-impact' or more discrete social issues to which they are better able to respond and achieve short-term results.

While the distraction paradigm seems plausible, and no doubt operates in certain situations, the mining company representatives I interviewed posited different reasons why companies tend to 'be distracted' from vital social issues. The value which multinational mining companies place on quantitative over qualitative data was foremost among these reasons. This preference constructs a linguistic, and by extension, practical divide between sustainability issues concerning economics, employment, the environment or governance and those issues related to societies and communities. The divide becomes particularly important when we consider that the internal decision-making processes which inform the types of community programmes implemented at operation sites are made chiefly on the basis of data presented to internal decision makers; data which is often the same as that used to compile sustainability reports.[11] Here, we see mechanisms of shared language and group identity in action: discourses and industry-wide cultural proclivities toward quantitative data influence issues-prioritization which is in tension with the idealized corporate values through which companies assert that they give equal weighting to social, economic, environmental and governance issues.

The role of comparable reporting: legitimacy through disclosure

Patterns in sustainability report content, similar styles of report presentation and the adoption of shared language position sustainability reports as a site where otherwise competitive organizations are largely united. Several pragmatic factors influence the similarities between different companies' reports, including: adoption of the GRI guidelines and related initiatives, such as the UNGC; the nature of annual reports and the necessity to make them comparable between years and between companies; and the growing presence of external auditing. In exploring the social mechanisms institutionalizing CSR at the industry level, it also becomes apparent that peer pressure is playing an important role in shaping how the studied mining companies define and approach their social responsibilities.

Again, sustainability reports provide a window onto these isomorphic pressures and the related quest for legitimacy. Through their sustainability reports, companies not only disclose their own CSR credentials, they also exert pressure for peer companies to do the same. The very act of reporting becomes one of both legitimation and normalization. The content analysis findings suggest that isomorphism and legitimacy are operationalized at the industry level in such a way as to promote a rather homogenous discourse and consequent practices. This process establishes shared performance benchmarks and, ultimately, creates agreed elements of legitimate social performance behaviours for the global mining industry.

As noted previously, one of the immediately noticeable shifts toward a shared language of responsible mining was the appearance of extremely similar report titles, with all companies eventually adopting some form of the title, 'sustainability report' by the end of the studied period. By 2006, four of the five companies also adopted language referring to their efforts to earn or maintain a 'licence to operate' from their stakeholders and communities. MMG's (2006) statement about this topic is representative of the three other companies which adopt the language of a licence to operate:

> The industry is recognising that its environmental and social performance is the key to maintaining its 'social licence' to operate. As a member of the Minerals Council of Australia, [MMG] is committed to the industry's 'Enduring Value: The Australian minerals industry framework for sustainable development'.

By their 2007 reports, all companies studied had adopted similar report layouts, all provided lists of the national and international sustainable development initiatives to which they are signatories or members, and the majority referred to their efforts to operate 'beyond compliance' or contribute to the 'triple bottom line'.[12]

In another example of the diffusion of shared language – and there are many – the reports uniformly adopt discourse of opportunities, achievements, challenges and targets. While a discussion of these basic concerns might seem obvious, what is interesting is that these diverse and competitive companies employ extremely similar language and styles of presentation as opposed to adopting differentiation strategies. Targets, for example, are consistently divided into environmental, social and employment issues and are presented in large tables, often using a 'stop light' format, using red, amber and green to indicate the extent to which aims were met or not.

From a new institutionalist perspective, the answer to why mining companies have adopted such similar sustainability reporting styles lies in the organizational legitimacy which may be derived from isomorphic change. While we easily recognise that multinational mining companies compete over access to ore bodies, supply chains and market share, their competition for legitimacy is just as real but more obscure. In the words of DiMaggio and Powell (1991), 'Organizations compete not just for resources and customers, but for political power and institutional legitimacy, for social as well as economic fitness.' Through sustainability reports, we can see the spread of readily accepted corporate social responsibility language and ideals (Strang and Soule 1998). The similarities which result, as evidenced through the shared language discussed above, signal the boundaries of and expectations for responsible mining, at least from the companies' perspectives. Moreover, these similarities build competitive advantage, through homogeneity (DiMaggio and Powell 1991). But what benefits are to be gained from the isomorphism through which mining companies' shared language on responsibility coalesces? And what does this suggest for responsible mining at an industry level?

Reflecting back on the discussion of 'targets' provides some insight into the benefits to be gained. This particular choice of language offers a utility for companies pushed to report publicly on areas in which they are likely to have negative impacts but in which stakeholders express interest. Instead of discussing failures, harm, shortcomings or negative impacts, companies are able instead to discuss targets for good behaviour and the 'challenges' they must overcome to meet those targets. This discursive positioning allows mining companies to acknowledge negative impacts without describing them as such and places them in a positive, action-oriented position in which they can discuss the way they plan to overcome those challenges in the future. Newmont (2005, 36) exemplifies the challenge-response style discourse common throughout all the companies' reports in its discussion of community relations employee training:

> One of the key challenges we face is the lack of capacity in the community relations area. A key aspect of our strategy, starting in 2006, is the development and implementation of a training program for community relations and development professionals.

Failure to report on negative impacts is a critique of sustainability reporting more generally. The GRI and KPMG (2008) note that stakeholders convey disappointment in companies' unwillingness to openly discuss negative impacts and their tendencies to report contributions and benefits in greater quantity, detail and depth than negative effects. The challenge-response approach used by mining companies reflects this trend and is linked to companies' risk management and marketing strategies. In a competitive and potentially litigious market, few firms would willingly print their shortcomings. Thus, despite its many critics, the challenge-response approach to sustainability reporting persists, suggesting that sustainability reporting in the global mining industry (or at least among the studied companies), may have crossed the threshold of innovation to enter the space where it offers organizational legitimacy instead of improved performance (Meyer and Rowan 1977). In other words, the shared language and peer pressures shaping the sustainability reports studied constrains innovative CSR practices within the industry, instead encouraging companies to mimic one another, lending credibility to select concerns while quelling consideration of other issues.

The shared language in mining companies' sustainability reports suggests the practice is moving toward a state of inertia. The point here is not simply that mining companies adopt similar language in sustainability reports but that the unified adoption of such language – fostered in part by the progressive implementation of the GRI framework during the years studied – lends legitimacy to all companies in the industry, even those which adopt less than rigorous reporting styles. This 'mimetic isomorphism' (DiMaggio and Powell 1991) reflects the uncertainties companies face when attempting to incorporate concerns for social responsibility more fully into their operations. By choosing to adopt language similar to one another, companies reinforce ideas about acceptable reporting styles, creating an industry-wide norm which helps dispel criticism. In the MCA's (2006) own words, the adoption of similar sustainability reporting styles across the industry:

> Create[s] significant reputational capital with local communities … differentiate[s] company performance … indicate[s] both good management

> and a lower risk profile for companies ... [and] has an increasing potential to become a significant factor in the ability of companies to attract investment.

Through this process sustainability reporting takes on a ritualistic formality which reproduces agreed elements of legitimacy, demonstrating that each reporting company is at least attempting to address those social, environmental and employment issues its stakeholders deem important. The noted change in the report space allocated to particular issues and the language used to discuss those issues may also indicate transformations in mining companies' social responsibilities that extend beyond the written page. Marc Jones (1999) theorizes that where a shared discourse gains prominence within sociocultural systems, that discourse significantly influences related on-ground practices. His assertion, based on a review of studies of institutionalizing pressures at societal, national and firm levels, also alludes to isomorphic pressures that influence companies to adopt language, policies and practices that echo accepted norms and legitimate resource allocations and activities across an industry. Such linkages are further supported through the work of 'voluntary regulation' in the form of industry initiatives, a topic we will pick up in Chapter 6.

Conclusion: aspirational CSR

The institutionalization of mining companies' responsibilities to the societies in which they operate reveals several components central to current policies and practices. These include: industry-wide values and standards evidenced through shared language; public communication of those ideals; and improved sectoral legitimacy through shared practice and group identity (Bice 2015). Together, these approaches contribute to a picture of responsible mining that can best be understood as 'aspirational'. Aspirational CSR works to publicly position mining companies as positive social agents attuned to their impacts and acting with the sustainability of affected communities in mind. In this aspirational mode, key tensions emerge from within, including: assertions about concern for social/community issues but a comparative lack of publicly available information on those issues; adoption of similar rhetoric, reporting practices and management approaches by firms which otherwise compete through market differentiation and domination; and support of voluntary reporting which sets out strong expectations for socially responsible behaviour but is engendered with little ability in terms of enforcement.

The analysis in this chapter makes clear that corporate social responsibility is institutionalized in the global mining industry. But the discursive positioning of community issues as the 'soft' side of the mining business within an industry which values hard science, masculinity and toughness suggests weaknesses. The progressive professionalization of community-related roles holds the potential to strengthen responsibility. It is this pillar of responsible mining to which we turn our attention in Chapter 4. Strengthening of CSR as an industry-wide institution may also occur as companies like BHP Billiton continue to assert a value proposition linked to non-financial responsibilities, even where quantitative data is lacking. While the 'beyond compliance' and 'social licence' discourses (Gunnigham et al. 2004) evidenced in sustainability reports and industry initiatives work to build legitimacy for companies' actions, these concepts must also mature in application and evaluation to maintain credence over time (Suchman 1995).

The aspirational CSR detailed in this chapter alludes to potential gaps between mining industry discourses and public disclosures and the more practical implementation of corporate social responsibility policies, management approaches and programmes at corporate headquarters and in communities. I turn to these meso- and micro-levels of analysis in the following chapters to further investigate how the social mechanisms influencing the adoption and perpetuation of CSR operate in practice and the implications this reveals for defining key principles for responsible mining.

Notes

1 As in previous chapters, the term CSR is problematic as companies use a variety of terms to refer to their environmental, social and governance policies and activities. CSR is used throughout this chapter as a catch-all term, for the purposes of discussion.

2 2003 represents the first year of sustainability report data analysed, but these reports were published in 2004. Similarly, most 2014 reports contain only 2013 data, although variations in reporting cycles mean some reports contain early 2014 data.

3 Pseudonyms and generic job titles have been used in line with the human research ethics requirements governing the research.

4 Pseudonyms are used for participants from a 2008–2011 study on mining companies' CSR policies and practices, in compliance with the University of Melbourne ethics agreement and participant consent forms governing this study.

5 For example, the 2007 Australian Federal Election was won by the Australian Labour Party largely on a platform which promised to address climate change and Australia's critical water shortage (Koutsoukis 2007).

6 Although Franks et al.'s (2010) guide to addressing the cumulative impacts of mining in the Australian context focuses on the coal industry, its central findings are applicable to various types of mining.

7 The Newmont 'Global Community Relationships Review' is considered a landmark study by many in the industry and the majority of senior managers I interviewed at corporate head offices mentioned or discussed the Review in detail. This quote does not, therefore, identify this individual as representing any particular company.

8 In my professional capacity as a CSR consultant, I wrote or aided in the production of companies' sustainability reports in a range of industries. In my experience, these reports go through at least three critical rounds of internal edits prior to publication, and all have been subjected to checks by legal advisors, senior executive approval processes and editing by marketing or communications departments.

9 During the time period studied in the content analysis, for example, BHP Billiton's full sustainability report ranged around 500 pages.

10 The sustainability reports analysed contain company data for the years 2003 to 2014. The report publication date tends to be one year out from the data covered but there is no consistent sustainability reporting period, unlike annual financial reports. For this reason, references may refer to bands of years (e.g. 2003–2004, 2013–2014).

11 In my experience of sustainability report writing, board papers comprise a key data source for report data, often with few substantive changes being made between end of year board reports and the information presented in the sustainability report.

12 In the case of MMG, the company went one step further, adopting a 'quadruple bottom line' which incorporated the now standard social, environmental and economic concerns, as well as governance.

References

AccountAbility. 2008. AA1000 Accountability Principles Standard 2008. London: AccountAbility.

Australian Government: Department of Industry, Tourism and Resources. 2006. 'Community Engagement and Development.' *Leading Practices Sustainable Development Program for the Mining Industry*. Kingston: Department of Industry, Tourism and Resources, 48.

BHP Billiton. 2005. 'A Sustainable Perspective: BHP Billiton Full Sustainability Report 2005.' Melbourne: BHP Billiton.

——. 2007. '"It's our…": BHP Billiton Sustainability Report.' Melbourne, BHP Billiton: 315.

Bice, S. 2015. 'Corporate Social Responsibility as Institution: A Social Mechanisms Framework.' *Journal of Business Ethics*: 18.

Blowfield, M. 2007. 'Reasons to be Cheerful? What we Know About CSR's Impact.' *Third World Quarterly* 28(4): 683–695.

Carolyn. 2009. Community Relations Senior Manager. Interview with S.Bice.

Catherine. 2009. Senior Advisor, Community Relations. Interview with S. Bice.

Centre for Social Responsibility in Mining. 2007. 'Assessing and Managing the Socio-Economic Impacts of Projects: A Review of Current Mining Industry Practice.' Brisbane: Sustainable Minerals Institute, 18.

Dashwood, H. S. 2004. *Norms Dissemination and Corporate Social Responsibility: An Assessment of Global Processes and the Global Mining Initiative*. International Studies Association annual meeting, Montreal.

——2007. 'Towards Sustainable Mining: The Corporate Role in the Construction of Global Standards.' *Multinational Business Review* 15(1): 47–66.

Deegan, C., M. Rankin and J. Tobin. 2002. 'An Examination of the Corporate Social and Environmental Disclosures of BHP from 1983–1997: A Test of Legitimacy Theory.' *Accounting, Auditing & Accountability Journal* 15(3): 312–343.

Deloitte. 2014. 'Tracking the Trends 2014: The Top 10 Issues Mining Companies Will Face in the Coming Year Canada', 42.

Diane. 2009. General Manager, Sustainable Development. Interview with S.Bice.

DiMaggio, P. and W. Powell. 1991. 'The Iron Cage Revisited: Institutional Isomorphism and Collective Rationality in Organizational Fields.' *The New Institutionalism in Organizational Analysis*. W. Powell and P. DiMaggio, eds. Chicago: The University of Chicago Press, 63–83.

Emberson-Bain, A. 1994. 'Mining Development in the Pacific: Are We Sustaining the Unsustainable?' *Feminist perspectives on sustainable development*. W. Harcourt, ed. London: Zed Books in association with Society for International Development, 46–59.

Ernst and Young and Boston College, Carroll School of Management Center for Corporate Citizenship. 2014. 'Value of Sustainability Reporting: A study by Ernst and Young and Boston College Centre for Corporate Citizenship.' Boston.

Franks, D., D. Brereton, C. Moran, T. Sarker and T. Cohen. 2010. 'Cumulative Impacts: A Good Practice Guide for the Australian Coal Mining Industry.' Brisbane: Australian Coal Association Research Program, 62.

Franks, D., C. Fidler, C. Brereton, D. Vanclay, F. Clark and P. Clark. 2009. 'Leading Practice Strategies for Addressing the Social Impacts of Resource Developments.' Brisbane: Centre for Social Responsibility in Mining, Sustainable Minerals Institute, 74.

Frost, G. and N. Martinov-Benny. 2010. 'Sustainability Reporting Assurance: Market Trends and Information Content.' *Non-Financial Information Disclosure Project*. Melbourne: CPA Australia, 18.

Garnaut, R. 2013. *Dog Days: Australia After the Boom*. Collingwood: Black Inc.

Genasci, M. and S. Pray. 2008. 'Extracting Accountability: The Implications of the Resource Curse for CSR Theory and Practice.' *Yale Human Rights and Development Law Journal* 11: 37–58.

Global Reporting Initiative. 2006. 'Pathways: The GRI Sustainability Reporting Cycle: A Handbook for Small and Not So Small Organisations.' Amsterdam.

——. 2010. 'Next Generation of GRI Guidelines on the Horizon.' www.globalreporting.org/information/news-and-press-center/Pages/Next-Generation-of-GRI-Guidelines-on-the-Horizon-.aspx (accessed 3 March 2016).

Global Reporting Initiative, The University of Hong Kong and CSR Asia. 2008. 'Reporting on Community Impacts.' Amsterdam, 33.

Gunningham, N., R. Kagan and D. Thornton. 2004. 'Social License and Environmental Protection: Why Businesses Go Beyond Compliance.' *Law and Social Inquiry* 29: 307–341.

Guthrie, J. and L. D. Parker. 1989. 'Corporate Social Reporting: A Rebuttal of Legitimacy Theory.' *Accounting and Business Research* 19(76): 343–352.

Guthrie, J. and K. Yongvanich. 2005. 'Extended Performance Reporting: An Examination of the Australian Mining Industry.' *Accounting Forum* 29(1): 103–119.

Hall, N., J. Lacey, S. Carr-Cornish and A.-M. Dowd. 2015. 'Social Licence to Operate: Understanding How a Concept Has Been Translated into Practice in Energy Industries.' *Journal of Cleaner Production* 86(0): 301–310.

Hall, N. L. 2014. 'Can the "Social Licence to Operate" Concept Enhance Engagement and Increase Acceptance of Renewable Energy? A Case Study of Wind Farms in Australia.' *Social Epistemology* 28(3–4): 219–238.

Hamann, R. 2004. 'Corporate Social Responsibility, Partnerships and Institutional Change: The Case of Mining Companies in South Africa.' *Natural Resources Forum* 28(4): 278–290.

Harvey, B. 2006. *Sociology Before Geology: The New Social Competencies of Mining*. 9th Kenneth Finlay Memorial Lecture, School of Mining Engineering, University of New South Wales.

Harvey, B. and S. Nish. 2005. 'Rio Tinto and Indigenous Community Agreement Making in Australia.' *Journal of Energy and Natural Resources Law* 23(4): 499–510.

International Council on Mining and Metals. 2014. 'ICMM Annual Review 2013: Strengthening relationships with communities.' London, ICMM.

Jones, M. T. 1999. 'The Institutional Determinants of Corporate Social Responsibility.' *Journal of Business Ethics* 20: 163–179.

Koutsoukis, J. 2007. 'Rudd Romps to Historic Win.' *The Age*. Morning edition.

KPMG International. 2013. 'International Survey of Corporate Responsibility Reporting 2013.' KPMG. London: SustainAbility, 36.

KPMG International and SustainAbility. 2008. 'Count Me In: The Readers' Take on Sustainability Reporting.' London: SustainAbility, 1–44.

——2010. 'Reporting Change: Readers and Reporters Survey 2010.' London: SustainAbility, 1–33.

Kunanayagam, R. 2003. 'Sex Workers: Their Impact on and Interaction with the Mining Industry.' Women in Mining Conference, 4 August 2003. 25.

Lahiri-Dutt, K. and M. Macintyre. 2006. *Women Miners in Developing Countries: Pit Women and Others*. Aldershot: Burlington Ashgate Publishing.

Meyer, J. W. and B. Rowan. 1977. 'Institutionalized Organizations: Formal Structure as Myth and Ceremony.' *The American Journal of Sociology* 83(2): 340–363.

Minerals Council of Australia. 2006. 'Inquiry into a Sustainability Charter: Submission no.94 to Parliamentary Standing Committee on Environment and Heritage.' Canberra: Minerals Council of Australia, 12.

Minerals Council of Australia. 2010. 'Socioeconomic Benefits and Impacts: An Assessment and Planning Toolkit.' Kingston: Minerals Council of Australia, 42.

Neuendorf, K. 2002. *Content Analysis Guidebook*. Thousand Oaks: Sage.

Newmont Mining Corporation. 2005. 'Now and Beyond 2005: Corporate Sustainability Report.' Denver: Newmont Mining Corporation, 68.

Oxiana [now MMG]. 2004. 'Sustainability Report: 2004.' Melbourne: Oxiana Ltd, 38.

——. 2006. '2006 Sustainability report.' Melbourne: Oxiana Ltd, 68.

PricewaterhouseCoopers. 2014. 'Mine 2014: Realigning Expectations.' 52.

Rio Tinto. 2009. 'Why Gender Matters: A Resource Guide for Integrating Gender Considerations into Communities Work at Rio Tinto.' Melbourne: Rio Tinto plc, 108.

——. 2005. 'Global Commitment with Local Solutions: 2005 Sustainable Development Review.' Melbourne: Rio Tinto plc, 45.

——. 2007. 'Sustainable Development Review.' *Rio Tinto 2007 Annual Report*. 83–94.

Schiavi, P. and Solomon, F. 2007 'Voluntary initiatives in the Mining Industry: Do they work?' *Greener Management International*, Issue 3: 27–41.

Strang, D. and S. A. Soule. 1998. 'Diffusion in Organizations and Social Movements: From Hybrid Corn to Poison Pills.' *Annual Review of Sociology* 24: 265–290.

Suchman, M. 1995. 'Managing Legitimacy: Strategic and Institutional Approaches.' *Academy of Management Review* 20(3): 571–610.

Susan. 2009. General Manager, Sustainable Development. Interview with S.Bice.

The International Monetary Fund. 2014. 'World Economic Outlook: Is the tide rising?' Washington, DC: The International Monetary Fund.

The World Bank. 2014. 'China Economic Update.' Beijing: World Bank Office, 31.

Walt. 2009. Head of Corporate Social Responsibility. Interview with S.Bice.

Williams, P., A. Gill and I. Ponsford. 2007. 'Corporate Social Responsibility at Tourism Destinations: Toward a Social License to Operate.' *Tourism Review International* 2: 133–144.

4 Holistic assessment and community-based agreement-making

The plastic table had seen better days. Warped and sun-worn, it wobbled precariously as I placed my laptop carefully to the left of a major crack. Wonky plastic chairs were scraped back as the three women and I settled in for what we expected to be a long discussion. 'Do you mind if I record our conversation? It will help me to make sure I reflect your ideas more accurately later.' I received three very polite assents and the women exchanged three subtle but clear 'here we go again' eyebrow raises. This was a relatively small, very remote developing country community, and the local mine had been operating for almost 20 years. Twenty years of change. Twenty years of impact assessments, research, negotiations and agreements. Twenty years of folks like me flying in to extract their data.

'Can I see the other ESIAs?' I asked. Although numerous environmental and social impact assessments (ESIAs) had been completed over the years, few were publicly available. The company's community relations supervisor pointed to a bottom shelf. 'They're all there, I think. But someone else had this office before me, so I'm still working out where everything's been put. But that looks like the lot.' I crouched and tugged a cinder block-sized report from beneath some folders. Grey dust settled back onto the shelf. The name on the front was well known and well respected in impact assessment circles. I flipped a few pages: demographic data, cultural heritage, language and dialect, gender issues, community maps, interview notes, summaries of corporate and government policies, discussion of a community complaints mechanism, budgets, livelihoods data, even a chapter on the area's climate and the ways in which this affected the community's movements and activities. 'Do you know what these have been used for?' I gestured to the other reports slumping on the shelf. 'I'm pretty sure that one was for the last ESIA,' my mining company host replied. 'You know, I've been meaning to but I haven't read it yet.'

Back at the card table, things had warmed up and the discussion was roaring along. Fuelled by cups of tea and the fortuitous drop-off of bread rolls, cheese and roasted chicken from a nearby catering oversupply, new women joined the main group for short periods, dipping in and out of the conversation, adding their stories to the mix. We had gathered around the card table to discuss the next iteration of the community's 'impact benefit agreement' with the locally operating company. The sweeping agreement covered issues from healthcare to education, law and order to support of youth and women's programmes and was valued at a small fortune in local currency. The original agreement took about five years to negotiate, the second iteration almost as long.

Women leaders from the community were taking an especial interest in this renegotiation. Previous agreements had incorporated benefits for women, but they had been kept away from the bargaining table, without a single representative signature. The investment of community members, company representatives, government and civil society groups in negotiating the agreement directly reflected its value to all parties. If done well, this community-based agreement could reap lasting benefits for local people in the form of building local business capacity, services and infrastructure. For the mining company, the agreement held the potential to reduce community conflict, especially considering a major mine expansion was on the cards with community resettlements pending.

Building a framework for responsible mining

This chapter begins to establish the five pillars for responsible mining: holistic assessment, community-based agreements, ethical decision-making, appropriate boundaries and good governance (see Figure 4.1). Over the course of the next three chapters we will draw on the context and conceptual framework blueprinted in the preceding chapters to articulate an approach to mining that integrates fundamental principles and practices with the aim of supporting a more socially responsible mining industry. The responsible mining framework reflects on what we have learned about the mining industry's core social challenges and socio-economic contributions. It accretes a variety of disparate practices, value frameworks, philosophical mindsets, social scientific findings and regulatory approaches to support a five pillared pathway that can better integrate social performance concerns into regular mining practice, now and into the future.

In this chapter, we will focus on the first two pillars: holistic assessment and community-based agreement-making. While the responsible mining framework is intended as an integrated and overlapping set of values and practices that can be applied at any stage of the mining lifecycle – from exploration to rehabilitation – holistic assessment marks a logical starting point. Wherever they operate, contemporary mining companies face requirements (to varying degrees of legal obligation) to test and predict their likely and possible social and environmental impacts. In many places, these requirements manifest in the form of social and environmental impact assessments. They therefore demarcate one of the more consistently engaged social performance practices. In disparate global jurisdictions, impact assessments (especially for environmental considerations) form a precursor to approvals and licensing. It is vital that all parties involved in the mining project identify and understand the probable or anticipated social, environmental, economic and governance impacts and benefits of any proposed, expanding or closing operation. When collected in a collaborative manner, with data publicly accessible, this information provides a shared starting point for the negotiations, ethical decision-making, relationship-building and work to follow.

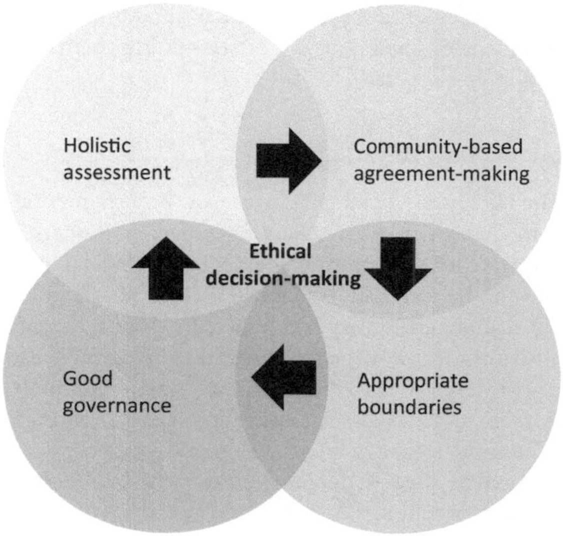

Figure 4.1 Responsible mining framework

Where government regulation may not require impact assessment, attention to 'non-technical risks' is now commonly required by major project funders, including via World Bank guidelines (Hatcher 2014), the International Finance Council's Performance Standards on Environmental and Social Sustainability or through major funders' commitments to the Equator Principles (Biermann et al. 2009). Defined as those environmental, social or governance risks outside the bounds of the geological or engineering risks to which companies traditionally attend, attention to non-technical risks serves to heighten delivery of impact assessments. As Professor of Human Geography and long-time SIA (Social Impact Assessment) pedagogue Professor Richard Howitt writes, 'Investment decisions require detailed prediction of outcomes before the commitment of corporate funds into new projects' (Howitt 2012, 81).

This chapter investigates and critiques the impact assessment processes mining companies commonly use to formally identify their social and environmental effects in the first instance. It discusses the history of impact assessment, pinpoints current shortcomings and advances concepts and methodologies to support more holistic impact assessments. The evidence and engagements possible through holistic assessment offer a strong starting point for responsible mining.

Community-based agreement-making builds on the strengths of holistic assessment. It offers an effective but surprisingly underutilized means of leveraging assessment findings to make informed decisions about balancing project costs with desired community benefits. The chapter goes on to explore how sound community-based agreement-making incorporates local knowledge and concerns while providing community members with opportunities to contribute feedback about and secure responses to ongoing impacts. Such processes can result in 'community impact and benefit agreements' (CIBAs) which aim to secure the meaningful, continuing involvement of local people in determining their futures, relative to the mining development process.

Impact assessment: current and past

Right now, somewhere, a social scientist, an environmental engineer, an anthropologist, a medical doctor, are collecting impact assessment data. Interviews are being conducted with community members, geological measurements are being taken, cultural heritage sites mapped, local health concerns recorded. Surveys are being completed in pen and paper. In Ghana, researchers are using iPads, with satellite internet allowing for immediate uploading of survey responses and transcribing of interview recordings. In Australia, community members are tweeting their ideas direct to locally operating companies, and in Europe they may soon be using Cambridge University-developed Q-cumber software on their smartphones to visualize proposed changes to local landscapes in 3D. Other software under development may allow community members to report fluctuations in a project's social licence in real time, and researchers have identified ways of plotting social networks through Bluetooth technology. In China, satellite communications and GPS allow medical practitioners to improve their health impact assessments through charting onto environmental issues in coal-mining intense areas. In Canada, crowd sourcing techniques and social media are helping First Peoples to contribute to cultural mapping, improving trust and data transparency. For the tens of thousands of impact assessments carried out each year, these techniques represent just a skerrick of the scads of impact assessment methods being used to capture community data.

Today's global application of impact assessment is rooted in 1970s America and the National Environmental Policy Act which required US Federal agencies to integrate social and natural science knowledge into decisions concerning government actions that could have an environmental impact (Becker and Vanclay 2003). Early development in the United States involved considerably broad stakeholder collaboration, setting the stage for more development-oriented impact assessment in the years to follow. The parallel rise of 'sustainable development' drew increasing government and corporate attention to social and environmental issues, further influencing impact assessment's advancement. Major environmental disasters of the 1980s, including Bhopal (1984), the discovery of the Antarctic ozone hole (1985) and Chernobyl (1986) cemented concerns for environmental impact and shaped a focus on non-financial issues (Elkington 1997). Environmental pollution and degradation moved onto the main agendas of intergovernmental bodies, with initiatives including the Brundtland Commission's *Our Common Future* report addressing growing anxieties about treatment of the environment (Elliott 2004). The Brundtland report famously connected environmental sustainability to intergenerational responsibility, intrinsically linking the environmental with the social. Partly as a reflection of this, most EIAs would today involve social impact assessment (SIA or ESIA)

– whether by requirement or proponent's preference – for capturing social data alongside environmental. This historical nesting of social within environmental achieves efficiencies but potentially undermines the value placed on social data, an issue to which we will soon return.

Concentrating on the social components of data gathering, the early days of SIA saw it deployed primarily as a predictive measure, intended as a planning tool and usually falling under related environmental assessment laws (Freudenburg 1986). This regulatory form of SIA (Nish and Bice 2012) positions it [as] 'a planning tool, prospective rather than retrospective – an attempt to foresee and hence avoid or minimize unwanted impacts' (Freudenburg 1986, 452). In practice, regulatory SIAs focus on anticipating social impacts and offering predictions about organic social systems from a static standpoint tends to concentrate efforts on completion of the assessment itself, not on SIA's real applicability and usefulness to communities over the long term (Ross and Lane 2001). As long-time impact assessment authority Professor Frank Vanclay writes, regulatory SIA offers 'no role for the management, mitigation and monitoring of impacts or for contribution of the SIA participants in the redesign of the project, or even in decision making about what constitutes an appropriate project' (Vanclay 2003, 2). Or, as Wolverhampton University Professor in Corporate Responsibility Michael Blowfield sums up, 'Companies frequently fail to identify their main impacts, so that the most significant consequences of the company's operations are often not the ones given greatest attention' (2007, 686).

Somewhat paradoxically, a predictive focus reduces regulatory SIA's ability to influence decisions and policies. This style of compliance-based SIA is today strongly criticized (Vanclay and Esteves 2012) and leading practitioners have long since steered away from assessments in which the means become the end, where the document created is viewed as exhaustive and final.

Even where compliance-based SIA is used to influence policies and decision-making, questions remain about its ability to identify the breadth and depth of a development proposal's impact on communities (Ross and Lane 2001). Especially for indigenous communities, regulatory SIA has failed to engage indigenous people or advance their ability to influence development in ways that favour their communities (O'Faircheallaigh 1999; O'Faircheallaigh 2012). A scarcity of full social impact awareness is partly to blame and can occur for several reasons, including: wide variations in data collection methods; limited or flawed data collection; weak theoretical linkages to social science; misdirected measurement tools; poorly trained practitioners; lack of auditing systems to check and validate assessment reports; politicization of the process leading to neglect of negative impacts; and limited incorporation of local knowledge or concerns to guide and inform the assessment (Burdge and Vanclay 1996; Lane et al. 2001; O'Faircheallaigh 2004). Prolific environmental sociology advocate Rabel Burdge, writing with Vanclay, adds to this list a tendency for regulators and corporations to adopt an 'asocietal mentality'(Burdge and Vanclay 1995). Such a perspective precludes individuals' abilities to understand and value social processes and social science methods and related findings, a step essential to extracting value from impact assessments. But, as leading researcher in mining and community relations Associate Professor Deanna Kemp suggested to me, partial responsibility for this situation also rests with social scientists:

> The fragmentation of the social sciences is part of that problem. Anthropologists, sociologists, demographers, social geographers, we have camps within an already small camp, so the social science voice is not collective. ... We're not able to collectively engage the industry because we come from such different standpoints ourselves, all of which have merit. But unfortunately, we don't stand collectively very well together, so we're not influential.

Industry perceptions of social scientists and their roles in impact assessment are inseparable from the style of holistic impact assessment vital to responsible mining, and this is an issue we will delve into later in the chapter.

Where impact assessment is driven primarily by regulation or transnational governance standards, it may fail to capture the desires of local landowners (Macintyre and Foale 2004). At mine-affected communities in Papua New Guinea, for example, this has played out through environmentalists' ideological construction of a 'noble primitive ecologist' – a false and patronizing identity which denies the concerns of local people while emphasizing sham dichotomies of nature/culture, other/western (Macintyre and Foale 2004, 235). Similarly, regulatory SIA may distance the assessment process from affected communities. This can result in community distrust of the assessment process or findings. Or, it may impose cultural expectations particular to the researcher onto assessed communities, as the time and repeat visits necessary to establish good working relationships never have the chance to form.

Nobody wants to have issues

The tendency for impact assessments to be viewed as outputs, better suited to life on a shelf than integration into practice may also relate to a very commonly held interpretation of 'impacts'. As one senior mining executive explained to me during a 2009 research interview to establish global criteria for social impacts, 'Nobody wants to have issues'.

Language matters. Chapter 2 detailed the importance of shared language to drive the adoption and perpetuation of CSR. It showed how shifting language reflects mining companies' changing understandings about their responsibilities to society – from a health, safety and environment focus to CSR to sustainable development, corporate citizenship and creating shared value. In Chapter 3, sustainability reporting language provided insights into how companies prioritize and discuss their non-financial activities. Here, language is central to how companies perceive their relationships with local communities and influences how they measure and report on their actual and potential impacts.

Returning to our conceptual framework, the process of companies carrying out impact assessments can be interpreted as a 'belief-formation mechanism'. The performance of impact assessments by multiple actors constructs its worth as a practice and influences its perpetuation (Hedström and Swedberg 1998). As such, it reinforces particular values that affect companies' understanding of their social impacts and responsibilities and, subsequently, how they respond to them. For example, regulatory ESIA prizes independence, a value which materializes in the form of strict requirements about how companies undertake the assessments and who can perform them. In practice, these requirements result in analysis and understanding of social impacts becoming decoupled from the on-ground activities intended to address them. According to one senior community relations manager of a leading miner, the decoupling that results from regulatory requirements for independent assessments 'mythologizes' certain practices vital to a company's social performance:

> There is a need for less consultant-dependent SIA. It's a tricky area because you need some expertise in relation to analysis, but too often the results from studies seem removed from regular practice; they seem mystical or masked in a kind of expert knowledge that seems unnecessarily inaccessible. This results, in part, from government provisos that require independence. So, the regulations themselves reinforce SIA practice that keeps the analysis process separate or formalized, relative to on-ground company staff.

Even where mining company employees are directly involved in the collection of information or discussions of what data to include in assessments, the ultimate authority for what is reported usually rests with the consultant. This may partly explain why many impact assessments are frequently treated as commercial in confidence documents, with data retained by the company. The practice of consultants collecting, analysing and reporting mining company data also creates distance between on-ground practice and the meaning of this practice at a more analytical level. If we take the sociological interpretation to a more esoteric degree – which may admittedly be a bridge too far for more sober readers – the ritualized handing-over of company data through ceremonialized reporting only reinforces the idea of SIA as a 'mystical' process.

The commonly occurring separation of on-ground community work from analysis of that work may contribute to companies' lack of understanding about priority community needs and, consequently, to ad hoc or reactive CSR programming. The act of assessment is an important step in understanding how a company affects communities and environments. The required 'outsourcing' of this step may inhibit companies' ability to reflect thoroughly on their social performance, community relations or sustainable development programming. Although many companies' community relations management approaches incorporate a traditional 'Plan, Do, Check, Act' method (Kemp 2009), in practice, 'checking' may be left up to consultants, with mining company employees expected to read and integrate the results into their practices. While such an approach seems reasonable on paper, there is something in the act of analysing a situation oneself which appears to be missed through outsourcing the task. A careful balancing of external expertise with company and external stakeholder involvement is necessary, and means of achieving this through a team-based approach to holistic assessment are introduced in the sections that follow.

Returning to the point that opened this discussion, assessments of socially related issues are also shaped considerably by the widespread discourse of 'impacts' to describe them. All mining company interviewees in my 2008–2012 study of Australian mining companies' CSR (see the appendices) stated that the term 'impacts' is perceived as inherently negative. During the course of that particular research project, this perception was projected onto questionnaires and interview instruments I used, despite my efforts to present a discussion about impacts in a non-judgmental manner. In one pre-interview questionnaire, for example, community health-related impacts were defined as, 'includ[ing] social effects related to diseases/illnesses or access to health care which may affect both the local community and the mine's operations.' Commenting on the questionnaire, one general manager said:

> My only comment would be that what you've captured tends to be some of the negative impacts as opposed to some of the positive ones. So that's just a comment and I suppose that's what you're looking at, but it's just that 'positive/negative' angle and what issues are being raised. Particularly when you talk about 'community health', I read that to be negative impacts: So it might be a gender issue, where you have greater transmission of HIV/AIDS with a male workforce, or those sorts of issues.

Comments like this emerged consistently when I discussed 'impacts' with interviewees. The negative discourse associated with regulatory ESIA, therefore, may also play a role in its ceremonial nature. Consequently, companies may unconsciously support the status quo of regulatory ESIA – which, at the most basic level, has been criticized for lacking meaningfulness or usefulness for both companies and communities – to avoid the more dynamic and frequent assessment of impacts which might occur under an ongoing ESIA regime like that recommended by the International Association for Impact Assessment (IAIA) and Frank Vanclay (2003).

Despite companies' reluctance to portray themselves as having social and environmental issues or impacts, compliance with regulatory ESIA persists throughout the mining industry. In this case, conformity to regulation outweighs concerns about negative discursive connotations and mining companies perpetuate the aura of evaluation to protect their formal structures (Meyer and Rowan 1977). Yet unhappiness with the negative linguistic connotations of ESIA remains significant. In Australia, for example, the Minerals Council of Australia (MCA) (2010) recently finalized its guide to 'Socioeconomic *benefits* [emphasis added] and impacts analysis (SEBIA)'. The language used to shape this guide veered actively away from the negative connotations of 'impact assessment'. In a workshop I attended, the consultant who developed SEBIA for MCA stated that mining industry members felt ESIA placed too much emphasis on risk identification and mitigation and not enough importance on social and environmental benefits to communities as derived from company activities (McMartin 2009). One of my interviewees, Susan[1] (2009), sat on the advisory panel for the SEBIA guidebook and she iterated industry concerns about the negative images conjured by 'impact assessment' during our interview:

> At the moment we're working on a working group with the Minerals Council of Australia to do a handbook to help people better understand impact assessment and that really sounds negative. So I think it's 'impact and benefit' and that's becoming more of the language because you can't get away from the fact that 'impact' sounds negative. There's just that perception around it, so you try to counteract that with the 'benefit' side of things.

Perceptions of impact assessment as a practice geared only towards outing the negative impacts of mining companies also stem from its common link with approval-specific mandates. Compliance-based impact assessment also carries values messages about which issues matter most. The tendency of regulation to Russian doll certain types of impact assessment within environmental assessments delivers a subliminal values message about the worth of social concerns (see Figure 4.2). As Deanna Kemp explains:

> We've seen some changes in systems: integrating human rights into social impact assessment, social impact assessment being more present in EIAs and so on. Things like agreements being part of the system of engaging [communities], but what we haven't seen is a structural change that acknowledges that mining is as much a social endeavour as it is a technical endeavour.

Except for a minority of mining companies, the potential contributions of impact assessment (IA) remain mostly contained within box-ticking approaches where, 'in *corporate planning systems* [emphasis in the original], SIA is most commonly mobilized as a technology for securing project approval' (Howitt 2012, 81). The persistent perceptions of IA as a means of regulatory threshold-clearing, however, are short-sighted and limit its potential. Former International Finance Corporation (IFC) advisor Nathan Monash suggests that better understanding and connection between technical and non-technical risk will play a major role in overcoming this particular standpoint. In an interview for this book, he echoed Kemp's perspective, saying:

> At a very local level the key is integrating non-technical and technical [risk]. Just to give a very specific example of that, I was working on a project where we were considering where to put infrastructure, and depending on the mine design you could have infrastructure that involved very long conveyors and tunnels. And that's fine. That may be your optimal design from a technical sense. But if you're in a very unstable area where such conveyors and pipelines are a target, and for political reasons may be under constant threat of being

> destroyed – and therefore you're going to have to have a massive security presence to protect that infrastructure if the current security [situation] continues – well that's not your optimal design anymore, is it? So bringing in security and political and community issues into mine design I think is critical.

For impact assessment, positioning social concerns under an umbrella of environmental regulation influences the ways in which social issues are approached and valued. From this perspective, social concerns are often seen as subordinate and 'soft', a perspective which is widely persistent in the mining industry (see Chapter 3) and one which hinders the extent to which socially focused impact assessments can benefit companies, communities and governments. Substantial repositioning of socially focused impact assessments as being valued in their own right and attitudinal shifts in management are necessary for this to occur. Otherwise – and much like the related situation in community relations roles – non-environmental impact assessment will remain 'core to business but not core business' (Kemp and Owen 2013).

Despite linguistic quibbles and substantial variations in the style and effectiveness of impact assessments for mine-affected communities, the practice retains considerable potential to promote 'democratic development' (Vanclay 2003). Howitt explains (without irony) that impact assessment is 'meant to change the world' (Howitt 2012, 78). And recent research suggests that the assessment process may be successfully interlinked with companies' CSR or social performance agendas (Bice 2015) and efforts to secure a 'social licence to operate' (Bice and Moffat 2014).

Where social, human rights or health impact assessments are valued as much as environmental, and where better integrated impact assessments are undertaken, all parties stand to benefit from a more comprehensive evidence-base for decision-making. This perspective creates opportunities for holistic assessments that adopt a systems lens to better capture and understand the ecosystem of impacts affecting any one project or community. For a responsible mining framework, the question then becomes: what will it take for the latent potential of impact assessment to be realized? Within a framework of responsible mining, what should impact assessment look like and how can it support the other four pillars?

These opportunities are within reach if we look to assessment methods at the leading edge of the field, especially integrated and cumulative impact assessment.

Before reviewing the methods available to advance holistic assessment for responsible mining, it is helpful to first reflect on how we came to be in the current situation. The above discussion reveals shortcomings in current industry practice and common perceptions of impact assessment, as evidenced through shared language. But we must also ask how impact assessment supporters and practitioners have failed to enact leading practice to achieve the levels of influence and positive outcomes in which supporters maintain strong belief. If we agree that responsible mining requires improved, more holistic assessment, and if we know that options from leading impact assessment practice are available, how can we explain the slow uptake of these methods within the mining industry?

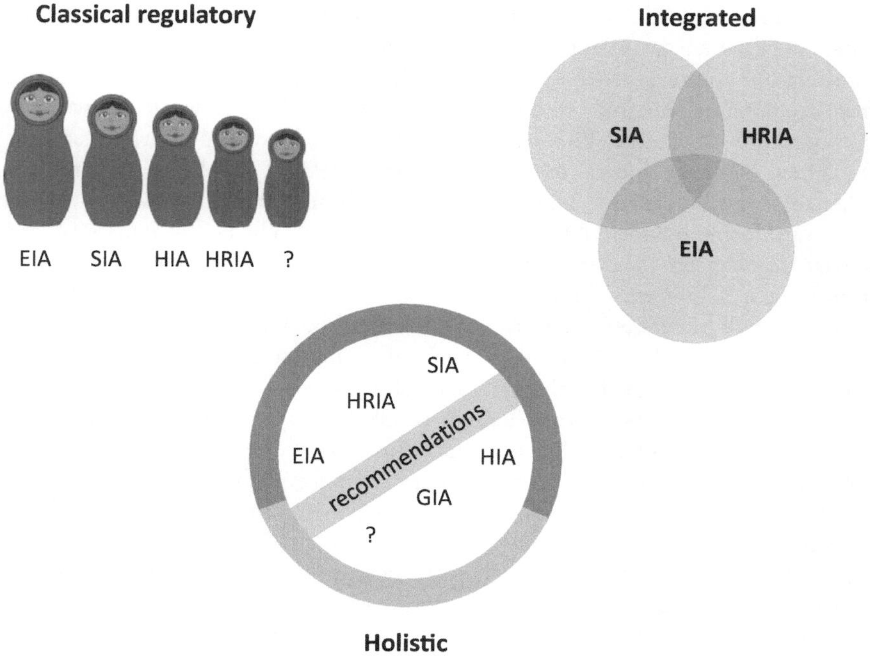

Figure 4.2 Approaches to impact assessment

From a social science perspective, a well-earned little side trip with one of the fathers of the discipline, Max Weber, followed by a reflection on the rationalized myth-making introduced in Chapter 2 may be instructive here. It is also necessary for the impact assessors and social scientists among us to turn the lens upon ourselves to consider our own culpability for current circumstances.

Weber helpfully gave social scientists the notion of the 'ideal type'. In the olden days, when social science was merely a nutty idea discussed over gritty espresso in a Heidelberg *kaffeehaus*, its founders were concerned that it be accepted as an empirical science. At the same time, Weber was a pragmatist and understood that boundlessly complex society combined with individual subjectivity would always hinder the ability of social scientists to establish a surety of results equivalent to those of their colleagues in white lab coats. To acknowledge and assist this problem, Weber argued for the importance of sound and rigorous social scientific methodology and for the use of ideal types for purposes of analysis. 'It is the task of sociology', he wrote in his 1922 *Collected Essays on Epistemology*, 'to reduce these concepts to "understandable" action' (Gerth and Mills 1946, 55).

The separation of impact assessments by various focus-areas could be interpreted as a natural extension of the original sociological project. On the one hand, topical focus encourages methodological robustness and empirical rigour. On the other, it allows for the parcelling out of ideal types of impacts and application of scientific (or pseudo-scientific) methods. Environmental concerns can be viewed next to social, alongside health or human rights, in a manner that arguably facilitates easier engagement with individual issues. Ideal types, however, mask impacts' interrelation and real-world complexities. They may also prevent research results from speaking meaningfully to one another. Howitt (2012) draws on the work of cultural critic Raymond Williams to suggest that SIA is 'an area where several disciplines converge, but in general do not meet' (1983, 15). This critique retains salience over 30 years later, and we will return to it in a moment.

Reintroducing the idea of the rationalized myth is also helpful in understanding impact assessment's shortcomings. Recalling the lunchroom model of CSR introduced in Chapter 2, we know that businesses rely on social norms to assist them to establish 'rationalized' rules to support their organizational structures and strategies. But we also learned that companies often struggle to bridge the gap between those rationalized myths and their real-world practice. In this way, corporate or government policies become loosely coupled or even fully decoupled from practice. These myths retain their acceptance and salience because it is only through collective buy-in to their existence that organizational equilibrium is maintained.

Like other CSR-related practices, impact assessment remains largely decoupled from core mining business. The instigation of Newmont Gold's global 'Community Relationships Review' is instructive here because it represents an outlier in which socially focused impact assessments garnered attention and commitment at the highest corporate levels and were actively acknowledged as core to investor concerns and corporate strategy.

In April 2007, the Newmont Board received a shareholder-led resolution 'requesting a report regarding Newmont's community policies and practices' (Newmont Mining Corporation 2009, 160). The resolution cited shareholders' concerns about a number of recent community conflicts and environmental degradation, especially in Ghana, Indonesia and Peru, contributing to project delays and fines. Newmont consequently made board-level commitment to undertake an 18-month-long review of community relations at its mine sites, globally. This unprecedented move marked one of the more successful shareholder accountability actions in contemporary mining history. Although the extent to which the assessments have been operationalized to inform company practice remains unclear, Newmont intended the study as an opportunity for corporate-wide review of community-related policies, standards, management and systems, making an explicit policy-practice connection. The review's uniqueness in the industry, however, is a stark reminder that assessment of social impacts lingers as a fairly marginal practice, usually contained within community relations or sustainable development departments and, where applicable, guided by regulation. Although legislative requirements and company policies may encourage a variety of impact assessments and establish their baseline importance, common practice has not yet advanced sufficiently to reduce policy-practice gaps.

Finally – and returning to an issue raised earlier – disjointed approaches to impact assessment may be steeped in the sociological past, but the continued siloing of the practice is answerable by contemporary social scientists. As global impact assessment nomads Angus Morrison-Saunders and Jenny Pope write with their colleagues: 'The proliferation of specialist practices has made it increasingly difficult to assimilate and integrate the [impact assessment] evidence in a single decision-making process in a way that is meaningful to decision-makers and which speaks of sustainable development' (Morrison-Saunders et al. 2014, 4).

In their impassioned call for more integrated impact assessment, these scholarly and practice leaders demonstrate that increasingly specialized impact assessment may undermine the relevance and utility of the very method they value. Even more importantly, they argue such disaggregated approaches stymie proponents' abilities to make long-term, evidence-based decisions. In their 2014 manifesto for integration of impact assessment to strengthen its influence and utility, Morrison-Saunders and colleagues delineate 46 distinct types of assessment (Morrison-Saunders et al. 2014). While each of these practices – from strategic environmental assessment to metabolic impact assessment, equity impact assessment to participatory impact assessment – offers its own punch by drawing upon niche expertise of practitioners, proponents and communities may suffer from too much assessment. 'The result of so much specialization', they write, 'is a proliferation of separate studies provided to decision-makers; multiple assessments leading to duplication of effort and inefficiency in conducting assessments; lack of integration and lack of focus, to name just a few' (Morrison-Saunders et al. 2014, 4). Multiplication of assessment topics and approaches also results in research fatigue, with substantial time and knowledge requested from communities and the likelihood of reduced participation rates.

So, how might impact assessment overcome a tendency towards ideal types, the loose coupling between policy and practice and the siloing of specialization? Holistic assessment is the key and emerging impact assessment practices may facilitate its development.

Emerging methods to support holistic assessment

Holistic assessment adopts a transdisciplinary and team-based approach to impact assessment to desegregate research silos, encouraging sharing of data, new styles of analyses and generating better connected recommendations for impact mitigation and benefit leveraging. It prizes three core characteristics: integration, integrity and dynamism. Holistic assessment profits from the advantages identified in interdisciplinary social scientific research, including: richness of data; clear interconnection of issues; understanding of values, competencies and power relationships influencing local situations; and the potential for novel and different insights through application of diverse perspectives (Flyvbjerg 2001).

At the University of Melbourne, early work in holistic assessment for resources related decision-making is in its preliminary stages. Through the Melbourne Energy Institute's (MEI) 'Sedimentary basin management initiative' geoscientists, agricultural specialists, water scientists, social scientists, legal specialists, economists, public policy scholars, engineers and technical scientists are combining their expertise to create an evidence-base for decision-making on sedimentary basin resource use. In Australia, this is no small task. Sedimentary basins underlie half the continent and the Australian community derives 90 percent of its primary energy from basin resources, mostly from fossil fuels (Rawling and Sandiford 2013). The basins also support the agricultural industry and supply rural populations with water. Basin resources are highly sought after, with a number of resources companies (including mining, oil and gas) already running operations or seeking approvals across states and territories.

Governments at Commonwealth, state and local levels have identified that information vital for decision-making about the management and governance of basin resources is lacking. Little is known about the cumulative socio-economic impacts of current resource usage choices. There is only limited understanding of how Australia's sedimentary basins work and how new technologies may affect precious water resources (e.g. Cook et al. 2013; National Water Commission 2010 and 2011). This situation led Australia's former Chief Scientist, Professor Ian Chubb, to declare 'the sustainable use of sedimentary basins' a Science and Research Priority for Australia. The holistic assessment being conducted by MEI responds to this challenge. Tasks currently being pursued (some pending funding) include three-dimensional computer mapping of the underground landscape; a planned roll out of seismicity monitors to create a baseline and measure induced seismicity from any resources or related activities; studies of carbon capture and storage opportunities; efficiency modelling of various resource use choices and technologies; pricing models; analysis of communities' opinions and protests via social media; industry and government engagement; study of the social licence to operate; policy analysis; stakeholder interviews and network mapping; and a plotting of relevant regulation.

The holistic assessment highlighted above is large and long term. It is unique in its situation within a major research institution and also through being driven by a focus on the resources in question, as opposed to being motivated by compliance or owned by any one proponent or agency. There is great hope that it will provide a model for future work in this area, but it remains a germinal process. Even so, it is beginning to realize the promise of a range of assessment techniques, especially through a commitment to deconstruction of silos and by drawing on methodological lessons from over 40 years of impact assessment. This includes advances in integrated impact assessment and the development of cumulative impact assessment.

Building on integrated and cumulative impact assessment

The idea of integrated impact assessment was introduced in the early 1980s in the United States and grew from a recognition that, even after only a decade of mandated EIA, benefits could be gleaned through confederated assessment (Rossini and Porter 1983). Integrated assessments encourage consideration of particular content areas from diverse perspectives, leading to 'interdisciplinary interaction' that reveals areas of overlap while retaining distinct approaches to the assessment process (Kauppinen 2012). The reduction of hierarchies between various impact assessment types and the placing of social, health, human rights and other 'soft' issues equal to environmental issues is a philosophical hallmark of the practice. In this way, it strongly reflects the values questions raised earlier. It also places great importance on stakeholder participation, a means of 'public accountability' that has 'a political and social value itself' (Bond et al. 2001, 1012).

Early work to establish an integrated impact assessment or 'integrated appraisal' was introduced at the turn of this century (Kirkpatrick and Lee 1999) but an agreed methodology remains under development (Bond et al. 2001). While this poses challenges, it is also reflective of the project-focused nature of impact assessments generally. This means that while assessment methods regularly fall into consistent categories, methods used vary greatly, with the most consistency visible between cost-benefit analyses or EIA methods. For example, the IAIA – the leading association for impact assessment scholars and practitioners – only introduced methodological guidance for SIAs, a long-practiced form of assessment, in 2015. For integrated assessment, research proposes that the approach adopted should: use consistent assumptions, methods and data; seek to avoid gaps but also overlaps; have all components undertaken at similar points in time; involve a wide range of stakeholders in a highly participatory way; involve the entire assessment team from the time of scoping to establish clear roles and perimeters for the entire assessment; value various components equally; and use interdisciplinary teams to interpret findings to produce overarching and comprehensive recommendations (Bond et al. 2001; Morrison-Saunders et al. 2014).

Cumulative impact assessment – the effort to capture 'the successive, incremental and combined impacts of one or more activities on society, the economy or the environment' (Franks et al. 2012, 202) – enhances the benefits of integrated impact assessment. It is an approach which, although rarely implemented, is well suited to the mining industry, as it is common for several mines to be in operation within a select geographical area. It acknowledges the complex ecosystem in which impacts occur and reorients the assessment from the perspective of any one company or project proponent towards a whole-of-community vantage point (Franks et al. 2010). Such a systems – as opposed to project-based – approach is evidenced in the MEI example outlined above. Cumulative impact assessment further incorporates more meaningful time horizons and employs forecasting about future developments and potential closures (Franks 2011). This is especially important for the community participation prioritized in holistic assessment, as this type of information is often held outside the public domain. Proponents of cumulative impact assessment also assert that its community-based perspective helps to overcome dusty shelf syndrome. Community members' involvement throughout the entire assessment cycle funnels into their ongoing role in monitoring impacts, providing input to strategies and collaborating to address identified issues (Franks et al. 2012). These recommendations imply a regular and dynamic assessment regime that becomes a shared means of engaging and addressing impacts, ongoing.

The holistic assessment introduced here enhances innovations from integrated and cumulative impact assessment by positing one overarching assessment where components are analysed 'simultaneously and in parallel', taking all perspectives into account while offering 'a joint stance on expertise and participation' (Kauppinen 2012, 345). This approach has the further benefit of encouraging results from divergent disciplines to be considered together, creating a clearer narrative of a project's total operating environment and effects. As such, it circumvents one of the major shortcomings of siloed assessments, a point reinforced by Finnish researcher Tapani Kauppinen (2012) in his work to establish 'human impact assessment', an integrated approach to social and health impact assessments:

> When [impact assessment] is conducted as an expert-orientated, shuttle diplomacy-style assessment, where experts use their own data, expertise and interviews ... the responsibility for evaluating and combining the different impact assessment results is normally left to whoever commissions the assessment and/or those actors concerned about the outcomes (citizens/stakeholders/ decision makers). The relevant actors would decide what resources should be allocated to each assessment approach, and how to interpret the results.

This comment raises a final, vital element of holistic assessment. Open data and transparent reporting are tenets of leading practice (Esteves et al. 2012) and are critical to the ability of holistic assessment to advance responsible mining. Siloed impact assessment places a great deal of responsibility on the commissioning organization – usually, the mining company – to collate findings from diverse studies and rationalize them in a valid and productive way. Holistic assessment, on the other hand, draws upon the expertise of the broad assessment team to analyse findings from divergent perspectives to construct a comprehensive set of recommendations. The shared creation of assessment recommendations not only helps to create a more thorough understanding of identified impacts and mitigation strategies, it also plays an important role in communicating results by providing a more complete narrative about project effects and benefits, externally and internally to the company.

Holistic assessment can also draw upon its interdisciplinarity to translate research findings into language that speaks to decision makers. As Nathan Monash explained to me, based mostly on his experience at the IFC, poor communication of issues related to non-technical risks can be a substantial hurdle to ensuring that the highest levels of company management understand their implications and the need to address them.

> This is a technical business, and we're [social scientists and community engagement specialists] on the non-technical side. It is not realistic – nor do I think necessarily appropriate that the rest of the business should learn to speak like we do or learn to understand these topics in ways that we understand – I think it's incumbent upon us to speak their language. And that's about dollars, schedules and risk. And so I think our ability to convert these things into those terms is critical to achieving them buying into what we're trying to do, what we all believe.

Holistic assessment faces the trials of an ideal type, and it requires commitment for substantive development. But the challenge is worth it, with a chance to enhance and extend the next wave of impact assessment techniques. Its potential is latent in Morrison-Saunders and colleagues' (2014, 6) musings on integrated assessment, where they write:

> [It] eliminates superfluous complexity; speaks in a unified voice to sustainable development and reflects good impact assessment practice that is purposive, rigorous, practical, relevant, cost-effective, efficient, focused, adaptive, participative, interdisciplinary, credible, integrated, transparent and systematic.

The preceding discussion established the importance of a dynamic, holistic impact assessment approach to responsible mining. But for the holistic assessment advocated as a central component of the responsible mining framework to be adopted, practical hurdles must be leapt over. The best assessment methodology remains impotent if no proponent is willing to fund it. And the right signals must be sent by government, demonstrating that holistic assessment will both meet compliance requirements and be supported in regulatory processes. As the United Nations Development Programme's Daniel Franks writes for the International Mining for Development Centre, 'a policy environment that encourages good social performance' is a vital precursor to strong practice (Franks 2012).

How might an environment supportive of holistic assessment be cultivated? Should the above arguments fail to convince, it is also possible to make a strong cost-based argument for investment in holistic assessment. Evidence suggests that the average EIA costs usually come in below one percent of total project investments and require from 22 days to four months of staff time. Given the size of projects to which assessment costs are related, however, the figures are substantial. In an average EIA in the US, costed at 0.2 percent of total project budget, the environmental assessment alone will cost approximately US$750,000 (Retief and Chabalala 2009). With several hundred EIAs expected to be lodged in 2015 in the state of Western Australia alone, global figures quickly escalate to considerable amounts. Additionally, recent research into environmental approvals processes in Canadian and Australian mining goes so far as to suggest that the processes constitute a business benefit, in terms of guiding investment decisions (Annandale and Taplin 2003).

If monetary and time cost-based arguments do not suffice, we can also turn towards holistic assessment's potential to enhance informed decision-making and reduce community conflict. Stakeholder networks are a critical component of the assessment process and community-based agreement-making – the second pillar of responsible mining – offers a considerable facility for bringing these opportunities together.

Community-based agreement-making: getting assessments off the shelf

If impact assessment practitioners had their own viral meme, it would inevitably feature a large but untouched report sitting sad and lonely next to the 1984 edition of the *Mining Almanac*, its white, all-caps impact font shouting, 'Ermahgerd! Imperсts!'. Even if we achieve the holistic assessment practices necessary for responsible mining, how can we ensure that the assessments are used in an active, regular and purposeful manner to advance responsible mining? How can we circumvent situations like the one that opened this chapter, where communities participate in yet another assessment cycle that is based on intentions of consultation and collaborative corporate-community relationships but which falls short of the promise?

For holistic assessment to provide a meaningful evidence-base for responsible mining practice, the reports must come off the shelves and enter into a dynamic interplay with community relations, strategic planning and regular business management. Community-based agreement-making (CBA) presents a field-proven but relatively underutilized opportunity to activate impact assessment findings, secure community participation and support economic wellbeing and local enterprise development (Nish and Bice 2012; Prno et al. 2010). It values a long-term perspective that employs assessment-derived evidence to collaboratively evaluate the balance of options associated with mining development. Research demonstrates that, when supported by a strong and openly accessible evidence-base, CBA can offer a dynamic process that reduces corporate-community power asymmetries, defines appropriate boundaries and roles, encourages community ownership of processes, issues and responses (importantly, this is not 'mutual responsibility'), prizes self-determination, and establishes long-term working relationships.

CBAs work very differently to widely entrenched compensation practices (e.g. royalties or compensation payments) and their consequent short-termism. As one corporate senior executive explained to me in an interview for this book:

> I think the temptation for mining companies is to say that you can, in a short-term, solve problems and generate goodwill by writing a cheque. Because, at the end of the day, the amounts of money – in the whole scheme of things – is not that great. But in the long run, you can create more problems for yourself.
>
> So I think that the best form of solution that comes up is the practice that's adopted widely ... around formal partnership agreements between stakeholders. And those would be typically brought into the camp of a community or government and the mining company, maybe other businesses, to sit down and – it's usually a lengthy process – to sit down to define what each party wants out of those circumstances and define what people are going to do and then to be bound in a formally legal way to live up to your commitments.

A clear understanding of the distinction between compensation and benefits is critical to CBA, and it is an area in which boundaries often become blurred. Resolution 88 Director and former Rio Tinto Global Practice Leader – Communities and Social Performance Bruce Harvey gave me his long-considered take on the issue, based on nearly four decades entrenched in community issues in mining:

> The compensation is recompense for something lost. ... Benefits are the things to be gained out of participating in the industrial or commercial venture. So, it means you have to actually join the activity in order to gain the benefits, because the benefits are engendered within the industrial activity.

> Here's an example. If the Ministry of Defence – and, in fact I think it did – if it acquired a pastoral lease somewhere in the Northern Territory 10 years ago as an artillery range, [then] that means by necessity that Native Title rights of access and hunting and gathering, [and] performance of ceremony are lost, because you're not going to be doing all that in the middle of an artillery firing event. And so compensation is owed for the loss of that right. But there're no benefits to be gained, because there's no economic activity engendered. There's no economic activity option coming out of that, right? It's pure loss.
>
> Now, if a mining company came in, and was looking at developing a giant, broad-acre mining operation over the same territory, it has the option to pay a benefit beyond compensation. The compensation it pays is exactly equal to what the defence department had to pay. It's the same loss. However, Aboriginal people of that land have every right to say, 'But we would like to participate in the benefits that accrue from the activity that you're about to undertake. So we're going to negotiate with you so that we can advance our joint enjoyment of the benefits.'

This example makes clear the socio-economic development opportunities inherent in mining projects. It links directly to CBA supporters' arguments that the practice facilitates a flow of benefits to affected communities that has historically been touted but rarely achieved (Prno et al. 2010) CBAs are therefore seen as progressive and mark an important shift away from a loss- or compensation-based company-community relationship towards one that opens greater opportunities for local communities to participate in the potential benefits of the mining operation. Their relative newness, however, means that comprehensive evaluation of effectiveness is lacking (O'Faircheallaigh 2004 and 2008a; Prno et al. 2010). Bruce Harvey again:

> You can be generous in the benefits space, rather than the compensation space. Beause technically, compensation is non-negotiable, isn't it? Once a government [says] 'Here's a schedule of values, this is what's lost, here are your payments'. ... And it just leads to everyone sitting on their backside waiting for rent. Now let's talk in this very constructive space of benefits. This other [option] is participating in the economic options of the future. So that's why it's a far better space to be working in.

At this stage in our discussion it is worthwhile acknowledging that for much of the preceding deliberation, communities appear as an object of corporate-led assessment and activity. While such a positioning has facilitated arguments core to this chapter, it belies the heterogeneity and agency present in mine-affected communities. As well-respected and deeply experienced resource industry consultant Ian Thomson told me in an interview for this book,

> It takes two to tango here. Communities [today] know more about what other communities get. They also know more about – and particularly in exploration – the real risk involved, and to some extent are saying, 'Well, you're probably only going to be here for six months, so we'd better get as much as we can and as fast as we can.' Demands have shifted up.

Community deliberation in the mining process raises the issue of free, prior and informed consent (FPIC). Focused on the rights of indigenous communities within their lands, the clear FPIC aims listed in the concept's name involve complexities of international law, human rights and performance standards. An entire volume could easily be devoted to FPIC alone. At its core, FPIC aims to ensure that communities take decisions without coercion, that decisions are taken based on full information, that no activities are begun without authorisation, that consent is actively sought and that communities maintain the right to grant or withhold consent for resources projects. For the responsible mining framework, the upholding of FPIC is implicit, and, where consent is granted, may be supported through community-based agreement-making.

The promise of community-based agreement-making

CBA developed first in Canada and is today used in numerous countries as diverse as Australia, Zimbabwe and Papua New Guinea. Although these agreements traditionally focus on indigenous communities (O'Faircheallaigh 2008b and 2013), the processes are applicable to local, mine-affected communities. They represent a step-change in mining companies' engagements with communities and may even signal a cultural turn in the way that mining companies understand their relationship to host communities. As Bruce Harvey described to me, community-based agreements evolved away from a (still very common) practice of compensation payments and community buy-outs. While he finds it more meaningful to use the term 'participation agreements' as opposed to the more widely used 'impact benefit agreements' (IBA) or CIBA, his comments make clear the intentions and advantages of CBA.

> We didn't like that idea of IBA, because it kind of lends itself to horse-trading, or trade-offs. So, some people came up with the term 'Participation Agreement'; we will jointly participate in each other's activity to the extent that we want each other to, and jointly participate in whatever future emerges out of our working relationship. So, that's where the Argyle [Agreement] got its name, the 'Argyle Participation Agreement'. I rather like that – that idea, you know, that we're all on this boat together. The more we learn to work constructively together and participate in each other's lives, the better off we're all going to be.

The Argyle Participation Agreement that Harvey refers to represents many of the positive qualities that make CBA the second pillar of the responsible mining framework. It shows that an agreed balance between benefits and burdens of future mining can be performed. And that it can be done through using an agreed and trusted evidence-base, with all parties committing to reconciling a difficult past while working towards a shared future.

The Argyle Diamond Mine is located in the East Kimberley region of Western Australia. A Rio Tinto-owned project, Argyle Diamonds commenced production in 1983, following the 1979 ore body discovery. The diamond deposit is substantial, with the mine providing the world's largest supply of naturally coloured diamonds. It is situated on Barramundi Gap, a cultural heritage site of particular significance to Aboriginal women. In 1980, Rio Tinto reached a limited agreement with certain Traditional Owners who consented to the destruction of Barramundi Gap in exchange for defined benefits. This limited agreement was rapidly seen as inadequate, both in terms of process and the benefits derived by the local Aboriginal community. It remained a source of ongoing tension within the community, with government and in the community's relationship with the mine for the following two decades (Harvey and Nish 2005). By 2001, and with the potential development of an underground mine extending mine life for a further 15 or more years, Argyle Diamonds committed to renew its relationship with local Aboriginal communities and enter into a new, comprehensive agreement process. In 2004, the Miriwung, Gidga, Malgnin and Wular Traditional Owners, Argyle Diamonds and the Kimberley Land Council signed the Argyle Participation Agreement. The Agreement was then registered as an indigenous Land Use Agreement under the *Native Title Act* in 2005 (Harvey and Nish 2005).

The Argyle Participation Agreement is founded on principles of mutual recognition, mutual obligation and active participation by Aboriginal people in decisions about mining and the benefits of the mining operation. Critically, the agreement – and all parties involved – remained staunchly committed to ameliorating the asymmetrical power relationships that plague company-community interactions. To achieve this, relationship boundaries and roles were clarified from the outset with considerable time allowed for working through a history marred by mutual miscommunication and failure to understand one another's perspectives. A formal apology for past wrongdoings was issued by Argyle Diamonds as a critical step to advance the negotiations (Northern Territory Social Justice Report 2006).

Reflecting the discussion on holistic assessment, all parties involved in the negotiation process worked from a shared information base. And it was one that incorporated not only the compliance-based information required of the company but also particular assessments requested by the community, including a community-controlled and endorsed ethnography that identified the current land-based governance systems, the range of indigenous interests in the mining lease and the relationship between those interests. Other information included: regional demographic profiles; baseline impact studies; impact assessments incorporating regional economic implications of Federal-State financial relations; social wellbeing indicators; company studies on the proposed style of block cave mining; a video tour of a similar mine in another state; and EIAs. Ensuring community trust in data was also paramount. To promote this, community members were involved in the EIA scoping process and Traditional Owners engaged a hydrologist to peer-review water studies. In an early act of citizen science – a practice we will come to in a moment – groundwater data was provided to Traditional Owners for their live monitoring.

The Traditional Owners' 26 member team was then facilitated to fully engage with this data through the support of independent advisors on issues ranging from legal to economic, financial to environmental. These experts assisted the community representatives to assess sociological, anthropological, socio-economic and other data provided to them by Argyle Diamonds. This process empowered owners to engage with data, translate it for their communities and deploy it in negotiations in a way that had not occurred in prior, unilateral interactions.

Regular and wide communication between all parties was prioritized throughout the agreement-making process. Importantly, the community established the process' logistics, timing and priorities, with a total of 74 (often two-day) meetings held during the negotiation period, resulting in five interim agreements before generating the final document.[2]

The Argyle Participation Agreement not only supported Traditional Owners to develop a strong governance capability for greater participation in the socio-economic benefits possible through the mining project, it also advanced the CBA process generally. Other examples of successful CBAs include the Western Cape Communities Co-existence Agreement (Australia), the Diavik Diamond Mine Participation Agreement (Canada) and a series of diamond mining-related CBAs in Canada (Prno et al. 2010). Within these processes we can see not only an emerging trend for more equal and robust company-community relationships but also support for community engagement in the impact assessment processes that support the agreements, especially in data collection and interpretation.

It is worthwhile taking a moment here to focus on the meanings and potential of community-owned data gathering and analysis for supporting responsible mining. This chapter tightly interweaves the first two pillars of the responsible mining framework: holistic assessment and CBA. And it does so purposefully, as the evidence-base on which decisions are made should live alongside agreements made. Although the motivations and processes for public participation in impact assessment processes remain multiple and sometimes divisive, benefits to communities are considerable where appropriate participation models are employed (O'Faircheallaigh 2010).Within these processes, research demonstrates that more positive results can be achieved where communities play a role in the gathering and interpretation of impact assessment evidence. As Natural Resources Institute scholars Jennifer Stewart and John Sinclair write, 'In practical terms, the benefits of public participation are numerous and touch on law, politics, conflict resolution, planning, and decision making' (Stewart and Sinclair 2007, 162).

Although much public participation in impact assessment remains *post hoc*, occurring through community meetings and public fora where assessment results are discussed, there is a growing call for more 'citizen science' wherein local community members become 'data owners' through their direct role in data collection. Processes like the groundwater monitoring in the Argyle case empower community members to understand their local situations more thoroughly while also building trust in data used by the mining company. As Nobel Laureate Peter Doherty skilfully articulates in *The Knowledge Wars* (2015, 63):

> The big problem with any highly specialised area of knowledge is that 'experts' can come across (often because of their own lack of communication skills) as authoritarian figures telling us what we should think. But we need to demystify

science and to suggest ways that anyone can check out a contentious area. Still, no matter how effectively the goal posts are moved from telling to showing, the ultimate truth of the matter is that science is as much a doing as a thinking game. In the end, being directly involved in something gives a much more immediate understanding than any amount of reading or being talked at.

Through holistic assessment and CBA processes, there is a distinct opportunity to leverage the technologies and techniques of citizen science. If Peter Doherty is right – and his past forecasting nous might suggest that he is – this will result in as-yet-unthought-of innovations and 'contribute to general, long-term wellbeing' (Doherty 2015, 68).

Conclusion

This chapter established the first two pillars of the responsible mining framework – holistic assessment and CBA. It posited several key points for optimal, holistic impact assessment to guide strong social performance. These traits include incorporation of a range of impact areas within an overarching assessment framework, capturing environmental, social, health, human rights, gender, cultural and economic concerns. It suggested that appraisals be undertaken in parallel with analysis occurring simultaneously within an interdisciplinary team to generate comprehensive insights and recommendations. Cumulative impacts may also be incorporated where relevant, and, at a minimum, holistic assessments can benefit from a community-led perspective, participation and collaboration on monitoring and impact management promoted in the practice. Research processes must be ethical and transparent, with data accessible to and understood by communities (at a minimum) with citizen science proffering a means of community ownership of data. Results must be translated for a corporate audience to effect change and be equally accessible to communities and government.

Holistic impact assessment can contribute to responsible mining, but like all other components of the framework, it must be utilized to its potential and integrated into the project lifecycle, with results informing daily practice. CBA was therefore presented as one workable way of achieving this. CBA offers a demonstrated means of ensuring that the evidence generated through good impact assessment is engaged to inform decision-making. Where communities achieve real bargaining power, their self-determination is realized, supporting viability over the longer-term, even after the mining project. CBA also holds the potential to shape company-community-government relationships to strike a healthy balance in roles and responsibilities, lessening problematic 'boundary creep', absent governments, community dependency and corporate overreach.

Impact assessment must be brought off the shelf. Reports that are technical, dense, take a long time to produce, are primarily anticipatory, not wholly inclusive of a heterogeneous community, are done solely by specialist consultants in multiple, niche areas, and lack community buy-in and ownership of data get us nowhere. Good impact assessment can get a mining project approved. Great impact assessment can change the future.

There is substantial work to be done in these areas to realize responsible mining. We need social scientists to coalesce, as a community of practice, if they are to gain respect and footing at the boardroom table. Asymmetric power relationships must be addressed and communities given real bargaining power. Companies must trust the ability of communities to engage in negotiation processes and to use their data not only in protest but in partnership. Equally, communities must be given the opportunity to unpick corporate data, to be involved in the collection processes and receive independent assistance to analyse results. Trust is yet again paramount. Evidence-bases and collaborative processes are foundational first steps, and they must be conducted through ethical decision-making, the pillar to which we next turn.

Note

1 This case is taken from a University of Melbourne study conducted between 2008 and 2012. The University ethics agreement guiding this study required that data be de-identified. Pseudonyms are used to protect the anonymity of respondents, in line with these requirements.

2 For a detailed discussion of this case, please see, Nish and Bice, 2012. The case study summary presented here is indebted to that work.

References

Annandale, D. and R. Taplin. 2003. 'Is Environmental Impact Assessment Regulation a "Burden" to Private Firms?' *Environmental Impact Assessment Review* 23(3): 383–397.

Becker, H. and F. Vanclay, eds. 2003. *The International Handbook of Social Impact Assessment*. Cheltenham: Edward Elgar.

Bice, S. 2015. 'Bridging Corporate Social Responsibility and Social Impact Assessment.' *Impact Assessment and Project Appraisal* 33(2): 160–166.

Bice, S. and K. Moffat. 2014. 'Social Licence to Operate and Impact Assessment.' *Impact Assessment and Project Appraisal* 32(4): 257–263.

Biermann, F., B. Siebenhüner and A. Schreyögg. 2009. International Organizations in Global Environmental Governance. Abingdon and New York: Routledge.

Blowfield, M. 2007. 'Reasons to Be Cheerful? What We Know About CSR's Impact.' *Third World Quarterly* 28(4): 683–695.

Bond, R., J. Curran, C. Kirkpatrick, N. Lee and P. Francis. 2001. 'Integrated Impact Assessment for Sustainable Development: A Case Study Approach.' *World Development* 29(6): 1011–1024.

Burdge, R. and F. Vanclay. 1996. 'Social Impact Assessment: A Contribution to the State of the Art Series.' *Impact Assessment* 14(1): 59–86.

Burdge, R. J. and F. Vanclay. 1995. 'Social Impact Assessment.' *Environmental and Social Impact Assessment* 30(1): 31–65.

Cook, P., V. Beck, D. Brereton, R. Clark, B. Fisher, S. Kentish, J. Toomey and J. Williams. 2013. 'Engineering Energy: Unconventional Gas Production.' Report to ACOLA. Melbourne: ACOLA.

Doherty, P. 2015. *The Knowledge Wars*. Melbourne: Melbourne University Press.

Elkington, J. 1997. *Cannibals with Forks: The Triple Bottom Line of 21st Century Business*. Oxford: Capstone.

Elliott, L. 2004. 'From Stockholm to Rio to Johannesburg.' *The Global Politics of the Environment*. 2nd edition. London: Macmillan, 7–28.

Esteves, A. M., D. Franks and F. Vanclay. 2012. 'Social Impact Assessment: The State of the Art.' *Impact Assessment and Project Appraisal* 30(1): 32–42.

Flyvbjerg, B. 2001. *Making Social Science Matter: Why Social Inquiry Fails and How It Can Succeed Again*. Cambridge: Cambridge University Press.

Franks, D. 2011. 'Management of the Social Impacts of Mining.' *SME Mining Engineering Handbook*. P. Darling, ed. Littleton: Society for Mining, Metallurgy and Exploration, 1817–1825.

Franks, D., D. Brereton and C. J. Moran. 2012. 'Cumulative Social Impacts.' *New Directions in Social Impact Assessment: Conceptual and Methodological Advances*. F. Vanclay and A. M. Esteves, eds. Cheltenham: Edward Elgar, 202–220.

Franks, D. M. 2012. 'Social Impact Assessment of Resource Projects.' *International Mining for Development Centre Mining for Development: Guide to Australian Practice*. Perth: IM4DC.

Franks, D. M., D. Brereton, C. J. Moran, T. Sarker and T. Cohen. 2010. 'Cumulative Impacts – a Good Practice Guide for the Australian Coal Mining Industry.' Brisbane: Australian Coal Association Research Program, 62.

Freudenburg, W. 1986. 'Social Impact Assessment.' *Annual Review of Sociology* 12: 451–478.

Gerth, H. H. and C. Wright Mills. 1946. 'Intellectual Orientations.' *From Max Weber: Essays in Sociology*. H. H. Gerth and C. Wright Mills, eds. New York: Oxford University Press, 45–74.

Harvey, B. and S. Nish. 2005. 'Rio Tinto and Indigenous Community Agreement Making in Australia.' *Journal of Energy and Natural Resources Law* 23(4): 499–510.

Hatcher, P. 2014. *Regimes of Risk: The World Bank and the Transformation of Mining in Asia*. New York: Palgrave Macmillan.

Hedström, P. and R. Swedberg, eds. 1998. *Social Mechanisms: An Analytical Approach to Social Theory*. Cambridge: Cambridge University Press.

Howitt, R. 2012. 'Theoretical Foundations.' *New Directions in Social Impact Assessment: Conceptual and Methodological Advances*. F. Vanclay and A. M. Esteves, eds. Cheltenham: Edward Elgar, 78–95.

Kauppinen, T. 2012. 'Human Impact Assessment as a Framework for Integration.' *New Directions in Social Impact Assessment: Conceptual and Methodological Advances*. F. Vanclay and A. M. Esteves, eds. Cheltenham: Edward Elgar, 341–354.

Kemp, D. 2009. 'Community Relations in the Global Mining Industry: Exploring the Internal Dimensions of Externally Oriented Work.' *Corporate Social Responsibility and Environmental Management* 17: 1–14.

Kemp, D. and J. R. Owen. 2013. 'Community Relations and Mining: Core to Business but Not "Core Business".' *Resources Policy* 38(4): 523–531.

Kirkpatrick, C and N. Lee. 1999. 'Introduction: Integrated Appraisal and Decision-Making (Special Issue on Integrated Appraisal and Decision-Making).' *Environmental Impact Assessment Review* 19: 227–232.

Lane, M. B., A. Dale and C. N. Taylor. 2001. 'Social Assessment in Natural Resource Management: Promise, Potentiality and Practice.' *Social Assessment in Natural Resource*

Management Institutions. A. Dale, C. N. Taylor and M. B. Lane, eds. Collingwood: CSIRO Publishing, 93–104.

Macintyre, M. and S. Foale. 2004. 'Politicized Ecology: Local Responses to Mining in Papua New Guinea.' *Oceania* 74(3): 231–251.

McMartin, S. 2009. 'Socio-Economic and Environmental Benefit and Impact Assessment (Sebia).' *Workshop on Innovations in Social Impact Assessment and Monitoring for the Resources Sector*. Brisbane: Minerals Council of Australia/Centre for Social Responsibility in Mining, University of Queensland.

Meyer, J. W. and B. Rowan. 1977. 'Institutionalized Organizations: Formal Structure as Myth and Ceremony.' *The American Journal of Sociology* 83(2): 340–363.

Minerals Council of Australia. 2010. 'Socioeconomic Benefits and Impacts: An Assessment and Planning Toolkit.' Kingston: Minerals Council of Australia, 42.

Morrison-Saunders, A., J. Pope, J. A. E. Gunn, A. Bond and F. Retief. 2014. 'Strengthening Impact Assessment: A Call for Integration and Focus.' *Impact Assessment and Project Appraisal* 32(1): 2–8.

Newmont Mining Corporation. 2009. 'Community Relationships Review: Global Summary Report: March 2009.' Colorado: Newmont Mining Corporation.

National Water Commission. 2010. 'The Coal Seam Gas and Water Challenge: National Water Commission Position Statement.'

——. 2011. 'On-shore, co-produced water: Extent and management.' *Waterlines Report Series* 54.

Nish, S. and S. Bice. 2012. 'Community Based Agreement-Making with Land Connected Peoples.' *New Directions in Social Impact Assessment: Conceptual and Methodological Advances*. F. Vanclay and A. M. Esteves, eds. Cheltenham: Edward Elgar, 59–78.

Northern Territory Social Justice Report. 2006. 'The Argyle Participation Agreement.' Darwin: Human Rights Commission.

O'Faircheallaigh, C. 1999. 'Making Social Impact Assessment Count: A Negotiation-Based Approach for Indigenous Peoples.' *Society & Natural Resources* 12(1): 63–80.

——. 2004. 'Denying Citizens Their Rights? Indigenous People, Mining Payments and Service Provision.' *Australian Journal of Public Administration* 63(2): 42–50.

——. 2008a. 'Negotiating Cultural Heritage? Aboriginal-Mining Company Agreements in Australia.' *Development and Change* 39(1): 25–51.

——. 2008b. 'Understanding Corporate-Aboriginal Agreements on Mineral Development: A Conceptual Framework.' *Earth Matters: Indigenous Peoples, the Extractive Industries and Corporate Social Responsibility*. C. O'Faircheallaigh and S. Ali, eds. Aizlewood's Mill: Greenleaf Publishing Ltd, 67–83.

——. 2010. 'Public Participation and Environmental Impact Assessment: Purposes, Implications, and Lessons for Public Policy Making.' *Environmental Impact Assessment Review* 30(1): 19–27.

——. 2012. 'Social Impact Assessment and Indigenous Social Development.' *New Directions in Social Impact Assessment: Conceptual and Methodological Advances*. F. Vanclay and A. M. Esteves, eds. Cheltenham: Edward Elgar, 138–153.

——. 2013. 'Community Development Agreements in the Mining Industry: An Emerging Global Phenomenon.' *Community Development* 44(2): 222–238.

Prno, J., B. Bradshaw and D. Lapierre. 2010. 'Impact and Benefit Agreements: Are They Working?' *Canadian Institute of Mining*. Ontario: Canadian Institute of Mining, Metallurgy and Petroleum.

Rawling, T. and M. Sandiford. 2013. 'Multi-Basin Usage/Cumulative Impact.' Background Paper for the Office of NSW Chief Scientist and Engineer. Melbourne: Melbourne Energy Institute.

Retief, F. and B. Chabalala. 2009. 'The Cost of Environmental Impact Assessment (EIA) in South Africa.' *Journal of Environmental Assessment Policy & Management* 11(1): 51–68.

Ross, H. and M. B. Lane. 2001. 'Social Assessment and Resource Management at the Australian Federal Level: Trapped in an Epistemological Corner?' *Social Assessment in Natural Resource Management Institutions*. A. Dale, C. N. Taylor and M. B. Lane, eds. Collingwood: CSIRO Publishing, 93–104.

Rossini, F. A. and A. L. Porter. 1983. 'Integrated Impact Assessment.'

Stewart, J. M. P. and A. J. Sinclair. 2007. 'Meaningful Public Participation in Environmental Assessment: Perspectives from Canadian Participants, Proponents, and Government.' *Journal of Environmental Assessment Policy & Management* 9(2): 161–183.

Susan. 2009. General Manager, Sustainable Development. Interview with S. Bice.

Vanclay, F. 2003. 'Conceptual and Methodological Advances in Social Impact Assessment.' *The International Handbook of Social Impact Assessment: Conceptual and Methodological Advances*. H. Becker and F. Vanclay, eds. Cheltenham: Edward Elgar, 1–12.

Vanclay, F. and A. M. Esteves. 2012. *New Directions in Social Impact Assessment: Conceptual and Methodological Advances*. Cheltenham: Edward Elgar.

5 Ethical decision-making

Weighing dilemmas, intuition and rational choice

Large-scale mining is nothing if not spectacularly technical and well-engineered. Yet previous chapters demonstrate that the mining industry must progressively engage with less technical and more intuitive challenges. Ethical dilemmas concerning business activities, human rights and the environment shape much of this work and may push mining companies into uncomfortable territory. Far from measured control, the field of ethics is instead rife with metaphor and interpretation. Dating back to the very roots of Western civilization, philosophy, myths and fables distilled universal principles and moral lessons retold today. These stories demonstrate that ethical dilemmas are regular to daily life but also demand a deep thoughtfulness and attention to far-reaching consequences.

Were we to choose an ethical fable for mining, we could do worse than to look to the Greeks' Pandora legend. In this tale, Pandora descended into the world, bearing her jar of wonders as punishment for human theft of the gods' fire knowledge. Her opening of the jar liberates hardships but also potential. Pandora's actions have long been associated with humankind's technical and mechanical competencies, and with the thrall to knowledge. Writing of the historical fascination with recreating life through machines, cultural historian Gaby Wood (2003) draws upon the Pandora myth. Feminist authors Sanna Karkulehto and Ilmari Leppihalme (2014) write of Pandora as representing 'the struggle between the will to knowledge and the repression of knowledge.' For mining, perhaps Pandora's Box is the earth, releasing opportunities and consequences at once terrible and necessary. Perhaps the earth itself is not the box, but instead Pandora's Casket freed the technical capacities to mine its riches. In either case, Pandora symbolizes the tension between our capabilities and the choices we make about how to use them. Certainly, for the Greeks, Pandora came to represent 'an element of their *own* [emphasis in the original] natures; culture founded on man-made things risks continual self-harm' (Sennett 2009, 2). Through Pandora, we see the decussation where the human capacity to exploit the earth's resources crosses risk. And this 'x' marks the home base for ethical considerations in mining.

The preceding chapter explored the exigencies of assessing mining's impacts in a holistic manner and creating participatory agreements with communities to address these impacts and leverage benefits. This chapter's concern with ethics is not far removed and further demonstrates the tangle of considerations facing the contemporary mining operation. At the abstract level, ethics in mining must deliberate the possibility of universal principles while preserving space for local distinctions. This requires skills in rational decision-making to yield ethical choices. At this more practical level, ethics in mining must contemplate business, human rights and environmental issues. Again, local values retain weight and must be emphasized in decision-making while upholding universal principles. For this reason, rights-based concerns are rising in prevalence in the industry, as evidenced through sector-based and transnational governance initiatives, especially the International Council on Mining and Metals (ICMM) *Sustainable Development Framework*, the United Nations Global Compact (UNGC) and the Ruggie Principles for Business and Human Rights.

In application, ethics manifest most commonly in mining company codes of practice and through signatory membership to industry bodies, as opposed to formalized regulation. The ICMM, for instance, requires members to 'Implement and maintain ethical business practices and sound systems of corporate governance' as the first of its ten Principles for Sustainable Development. The ICMM does not, however, go further to delineate what such 'ethical business practice entails'. Indeed, many historical ethical dialogues encourage deliberation on a case-by-case basis and argue against universal 'answers'. As former EITI Board member and extractives industry policy expert Eelco de Groot explains, the positioning of ethical initiatives within the mining and extractives industry inclines towards voluntary, transnational governance initiatives. He also notes that communities affected by mining are now more aware than ever of the style and degree of treatment and compensation other mine-affected communities receive, placing further pressure on companies to act ethically. As de Groot puts it, 'The world is now very connected, and this becomes very important. … It has this beyond compliance nature. [It's] international ethics and what "the other" has got. And that's basically the felt human fairness, and not your national legislation.'

While explicit consideration of ethics in the mining industry remains a relatively new area of inquiry, contemporary codes of conduct, transnational governance principles and even the integration of employee-management accountability instruments, like Newmont's use of the 'EthicsPoint' online 'Ethics Solutions Tool', belie perennial metaphysical deliberation. Following this tradition, this chapter is a meditation on the relationship between corporate responsibilities and ethics, and on the particular positioning of their intersection in the mining industry. It canvasses core ethical considerations in three key areas – business, human rights and environment – to discuss the variety and importance of competing ethical demands on the contemporary mining company. It engages ethical scholarship to suggest the perspectives and techniques that may be most effective to support ethical decision-making processes and suggests ways in which mining companies can tap existing skill sets to improve their ethical capabilities.

Universal principles, local diversity and rational decision-making

Ethicists and philosophers have debated the gazetting of ethics for centuries. Max Weber writes of the challenges of extricating morals from ethics in the early seventeenth century, a time when both were inextricably entwined with religious dogma. His points may register as esoteric, but his concerns were linked directly to the role that ethics and morals play in the taking of decisions in everyday life. For Weber, the elucidation of universal ethical principles 'was connected to the idea of the beyond', 'ideal types' which would facilitate 'good' decisions in a 'logically consistent form' (Weber 1992 [1930], 68–69). The appeal of clear ethical principles to business sense emerges even here; universal ethical principles offer an assured and efficient logic to inform decisions that support moral conduct. The work of 18th century philosopher Immanuel Kant retains similar relevance for contemporary ethics, focused as he is on reason and rationality. When Kant asks, 'Is it a necessary law *for all rational beings* that they should always judge of their actions by maxims of which they can themselves will that they should serve as universal laws?' (Kant 1785 [1989], 55), he invokes the reasonable person and presents a position point from which to define ethical principles. It is not too great a stretch from Kant's imperative to ask: is there a universal ethics for the global mining industry? And how can multinational miners attend to the potentially competing demands of business, human and environmental ethics? To what extent can ethics be reasonably incorporated into corporate decision-making? To what extent can companies afford not to enmesh ethics in their deliberations?

Ethical debate has arguably intensified in light of multinational corporations becoming subject to transnational governance, an issue we will take up in Chapter 7. Here, it is helpful simply to bear in mind two contexts. First, that the debate concerning ethics is longstanding and will continue. Second, that contemporary ethical considerations sit in close proximity to shifts in responsibility resulting from globalization and consequent transnational mining governance. When globalization sociologist Saskia Sassen tackles questions of governance and responsibility, she raises the relationship between leadership and obligation, saying, 'transnational institutions and regimes do signal a shift in authority from the public to the private when it comes to governing the global economy' (Sassen 2007, 225), citing environmental protection in large projects as an example. Such authority brings with it expectations to consider the reasonable and even universal principles through which action should be scrutinized. Questions of appropriate boundaries, defined through corporate and government roles and responsibilities are of equal import, and are interrogated in Chapter 6.

At the same time, strong arguments are put forward to temper a dogmatic appeal to universal principles. Business ethicists Campbell Jones, Martin Parker and René ten Bos acknowledge that universal ethical principles can provide helpful guidance but caution against accepting such principles as a strict solution (Jones et al. 2005). Instead, they make the case for a consistent willingness to deal with each dilemma on its merits to allow for incorporation of diverse social, cultural and political perspectives; that which philosopher Emmanuel Levinas captures in the notion of 'the Other' (Levinas 2006). Local considerations of 'moral accountability', power, interest, knowledge-production systems and authority represent further, critical and specific aspects of ethical deliberations (Owen 2005). These are impossible to capture in static, universal frameworks, much like our middle range theorizing in Chapter 2. In this way, ethics for the global mining sector necessitate a considered balance. They require appraisal of universal principles tempered by expertise in local contexts and a prudent ability to discern how to weight and respond to each in any given dilemma.

The responsibility for global miners to understand and adopt ethical ways of thinking and doing poses a substantial challenge. In many ways, good ethical practice is accomplished through the very processes that make it difficult to define conclusively and universally or to systematize. In other words, it is the constant questioning which heartens ethics. And it is why Lawrence Kohlberg's theory of moral development (Kohlberg 1984) – wherein individuals progress in their ability to consider the positions and concerns of others through experience and responsiveness, with those most developed appealing to universal justice – retains prominence in the teaching of business ethics (Furman 1990). Questioning and experience lead managers to 'cognitive sophistication' of the type that facilitates ethical decision-making. Thus, the web of ethical issues associated with the global mining industry must be seen from beyond the corporate enclave – it involves the positions, choices and values of individuals, corporations, local communities, firms in non-mining sectors, governments, consumers and the general public.

Such holistic considerations are well and good. But they presuppose a business environment that supports the often abstract rationality and reasoning required for ethically based decision-making (Furman 1990). This is a critical consideration. While a detailed discussion of ethical decision-making processes is beyond the bounds of this chapter, it is worthwhile briefly contemplating their place and potential in the mining industry. Although the sector might seem uncongenial to the abstract requirements of ethical reasoning, the industry is driven by a technical rationality and precision not dissimilar to the rational contemplation demanded by subjective and imprecise ethical dilemmas.

While ethics cannot be equated with a wholly unfeeling rationalism – something deduced through an almost mathematical analysis – ethical decision-making does demand rational thought of a particular style. For example, such rationality requires the ability to view complex issues from varied perspectives and to evaluate the 'moral intensity' of the issue in question (Jones 1991a). Decision makers must attempt to understand how individuals or groups in different positions rationalize their own 'right' answers. One experienced vice-president interviewed suggested that employees can be taught how to integrate such decision-making methods into their daily work through targeted education programmes, thereby embedding ethical considerations within the mining company. 'I think ethics training is so important in that it gives people a framework to think about problems from a number of different perspectives. And it doesn't really matter which perspective you end up adopting. It's the exercise of thinking about things in a different way which is important. And if you can get a group of mining executives to think about problems in a different way, you'll almost always come up with better solutions than you would have otherwise.'

For philosopher John Rawls, this thinking exercise is possible solely behind the 'veil of ignorance' where impartiality is the only position from which ethical choices are possible (Rawls 1971). Or, as Max Weber suggested, rationalism 'embraces a world of opposites' (Weber 1992 [1930], 27). While the rationality required to run a successful mining operation may be applicable to ethical deliberation, it must also, therefore, be tempered with local values and subjective considerations.

At the abstract level, the ethical tangle for the mining industry is one threaded with appeals for universal ethical principles, the diverse values and positions of dissimilar communities, and companies' capabilities for ethical reasoning. According to certain research, a company's ability to unsnarl ethical dilemmas is strongly influenced by corporate leadership, especially conformity to ethical ideals and related accountability (Wood et al. 2006). As Greg Thompson, the long-time Executive Director of Transparency International Australia noted, in an interview for this book, ethical leadership is a corporate characteristic gaining in value. 'This is the competition now, to be seen to be the best company because of the ethical quality of the leadership of the company'. At the day-to-day level, ethical decisions must be achieved while ensuring or even boosting business viability. This situation introduces the critical intersections between a capacity to extract resources and human and environmental risks, noted earlier.

The demands and consequences of ethical business behaviour are not insubstantial. And it is an area which can be highly personal. Ethical decisions rely heavily on the judgement of individuals within the firm and on the tone and expectation set by internal leaders (Wood et al. 2006, Global Business Citizenship). As policy consultant and former BHP Billiton Vice President for Business Conduct and Public Policy, Holly Lindsay explains, executive values are a critical influence on ethical corporate behaviours, but these responsibilities extend throughout the firm. 'I think the CEO [and] the corporate culture are very, very important, because the CEO has the capacity to change the culture. Of course, it takes some time to do so, particularly in very large companies. It takes a long time to change a culture. So tone at the top is very, very important; but tone in the middle is also very important.' Additionally, as Oxfam Australia's Mining Advocacy Lead Serena Lillywhite says, strong individual judgement is powerful but only within appropriate, supportive systems. 'Within that ethical piece, especially with regard to honouring commitments, that is only going to happen if you have a robust system of checks and balances. The counter balance to transparency is, of course, accountability and in some way, it's about accountable relationships. That's what seems to be really missing: the monitoring, the accountability'.

Considerations like the above present yet another entry-point into the debate about CSR's business case. They also admonish us of the ease with which considerations of ethics or values can quickly be reduced to cost-benefit style appraisals. Infamous cases like that of Ford's Pinto – in which strict rationalization via cost-benefit analyses led to the loss of human lives (Gioia 1992) – remind us of the perils of over-rationalization. Theoretical efforts to link CSR and ethics further support this point. In their influential 'social contracts' approach to business and society relationships, for example, Thomas Donaldson and Thomas Dunfee (1999, 6), remark on 'hard' (universal) and 'situational' (based on 'pure' rationality) ethics:

> Ethics has been a subject that people have likened to stone; ethics must be rock-hard, they assume, or "situational" ethics may intrude. 'Situational' or 'conditional' ethics, in turn, popular wisdom has it, lead to the worst sort of ethical rationalization. Someone who is honest when it is convenient to be honest, and dishonest when it is not, is said to be morally bankrupt.

At the most philosophical level, ethics present mining companies with intricacies and tensions that are simultaneously vital to consider but which also require contemplation of ideas more obtuse than operational concerns. And there are no easy answers. As Donaldson and Dunfee point out, it is even difficult to take a position on hard vs. soft ethics, since, 'two conflicting conceptions of ethics can sometimes both be valid, and also in the sense that community agreements about ethics often matter' (Donaldson and Dunfee 1999, 6–7). Other scholars lament the lack of a 'ready to wear' business ethics model (Wood et al. 2014). Positioning ethics as a core component of 'global business citizenship', Donna Wood and co-authors (2014, 46) articulate the universal principles/local values tensions faced by multinational corporations:

> A global business citizen accepts a limited number of basic universal principles, such as, 'it is wrong to harm innocent persons.' However, in conducting business activities, this organization realizes that although application of the fundamental principles is straightforward in many cases, there are situations where local norms appear to be in conflict with those principles, or application of the principles will cause unintended negative consequences. There are even situations where the local manager cannot tell whether local customs conform to or conflict with company norms, or whether the comparison is even relevant. In these cases, the degree of ethical certainty is much lower.

In other words, negotiating ethical uncertainty poses risk for responsible mining company behaviour.

CSR is not ethics; ethics is not CSR

Corporate social responsibility is many things but it is not ethics by another name. The distinction, however, is subtle, and it is instructive to attempt to tease them apart. Many theorizations of CSR are underpinned by a central assumption that firms must consider social or ethical factors other than profit motive in decision-making and in action (Bird and Velasquez 2006). Professor Michael Blowfield, for instance, sees the CSR-related moral dilemmas linked to the challenges of supply chains, labour, and ethical sourcing of natural resources as *drivers* for ethics (Blowfield 2004). Here, CSR is viewed through an ethico-political perspective which sits opposite to business case perspectives (Hanlon 2008). Professor of Business Ethics David Vogel, meanwhile, distinguishes between complementary areas of business ethics and corporate social responsiveness to separate the values underpinning decisions from actions (Vogel 2005). Approaches like these are commonly classified under an 'ethics' label, distinguishing a particular taxonomy of CSR theories (e.g. Garriga and Mele 2004; Klonoski 1991; Windsor 2006).

Ethical approaches to CSR place considerable onus on the corporation to understand and enact norms, including through judgement encompassing organizational and local values (Kuhn and Deetz 2008). These approaches tend to be quite broad, invoke appeals to universal human rights, and imply that corporations must take on roles which may rest outside of their historically defined remit in order to achieve social good, a positioning affecting the mining industry in ways we will explore in Chapter 6. A survey of ethical CSR definitions reveals several other shared principles. For instance, there is widespread agreement that an ethics-based approach to CSR requires firms to look outside concerns with their financial bottom lines and beyond minimum obligations to law, shareholders and standards (see Figure 2.1). CSR guru Archie Carroll (1999), for example, suggests that corporations must be aware of and responsive not only to legal obligations, but to economic, ethical and 'discretionary' expectations of society. Working 'beyond compliance' is a well-accepted idiom in mining and in other natural resource industries (Gunningham et al. 2004). Michael Hopkins (2007), whose work focuses on the contemporary role of companies in fostering socio-economic development, asserts that firms must extend themselves beyond primary operations in order to improve living standards for internal and external stakeholders while conserving financial viability. Other approaches couch ethics as a foundation stone for CSR, linked closely to firm legitimacy and employee behaviour and morale (Aguilera et al. 2005), (see Table 5.1, below).

Table 5.1 Common scholarly approaches to CSR (adapted from Bice 2015)

Defining traits	*Business case models*	*Stakeholder models*	*New institutionalist models*	*Ethical models*
View of the corporation	Profit-focused firm which exists to create shareholder value	Corporate citizen	Multi-layered 'agentic' actor influenced by social norms and pressures	Moral agent whose collective conscience reflects that of individual executives/ managers
Theory of change	Firms increase profit, improve attractiveness to investors and create long-term financial viability through attention to social, environmental and governance concerns	Through establishing strong stakeholder relationships, firms achieve legitimacy or earn a social licence and thereby forestall protests or outrage	Firms' actions, at multiple levels are shaped by and shape a complex web of highly integrated social drivers, with CSR as a social process contributing significantly to firm longevity, relationships, ethical reputation and social roles	Firms bear a moral responsibility to global society and the firm may become threatened or the moral identity of individuals within the firm sullied if this is not upheld

Ethics is the sheep in a box

We have thus far determined that ethics in the global mining industry invokes universal principles attuned to local values and requires particular skills and mindsets for decision-making. We might also agree that CSR is not ethics and that ethics is not CSR. So, what is ethics?

'If you please, draw me a sheep!' So demands Antoine de Saint-Exupéry's Little Prince of the weary pilot in the classic children's fable. The pilot, an erstwhile artist, puts forth his best drawing, to which the Prince responds, 'No, no, no! I do not want an elephant inside a boa constrictor!' The pilot, eager to comply, draws several other versions of sheep, each rejected in turn for various shortcomings. Finally, in a last-ditch appeal, he draws a small box with air holes. 'This is only his box. The sheep you asked for is inside', the pilot explains to a delighted Little Prince. For philosopher Emmanuel Levinas, ethics is the sheep in a box. 'I do not know how to draw the solution to insoluble problems', Levinas writes (Levinas 1967 [1999], 89). Do ethics pose an 'insoluble problem' for the global mining industry? Can an industry which prides itself on technical capability, efficiency and problem-solving engage successfully with more subjective and mutable ethical dilemmas?

The answers to these questions seem, rather paradoxically, to be 'yes' and also 'yes'. Ethical dilemmas will necessarily proliferate in an industry that requires substantial trade-offs between resources and the environment, demands the cooperation of diverse communities, poses dangers due to the nature of work (even in the most safety-conscious of workplaces) and that involves utilization of a finite resource. These problems are not 'solvable', in the purest sense, but they can be workable, and ethical considerations comprise an important part of responsible solutions.

Mining industry initiatives guiding ethics

We know that much of the work to address ethical issues in the global mining industry has occurred through voluntary codes of international bodies, performance standards and company-specific codes of conduct. Certain of these standards reveal the overlap between ethical codes and transnational governance in that they provide both broad principles for responsible mining business practice and require that certain standards are met and reported on. We will explore this intersection further in Chapter 6. For now, what does the global mining industry currently say about ethics? And what are the most influential ethical guides and codes directing the industry's moral compass?

The recent proliferation of voluntary codes to inform responsible business practice makes it unwieldy to scrutinize their myriad. But we can pin down a few commonly used guidelines and survey their core contributions to inform ethical decision-making for the global mining industry. These guidelines play an important role in assisting the industry to define and attend to ethical concerns, representing a notable expansion in business priorities. As Transparency International's Greg Thompson notes, 'It's not just about numbers anymore. It's about what lies behind those numbers, what decisions have been made, what drives those decisions. Ethical questions then become more central.'

Business ethics

The ICMM, whose history is touched on in Chapter 2 and explored in fascinating detail in Daniel Franks' *Mountain Movers* (2015), has been instrumental in accentuating ethical concerns for the global mining industry. As noted, their ten Principles for Sustainable Development explicitly address business ethics in Principle 1: 'Implement and maintain ethical business practices and sound systems of corporate governance'. In 2012, ten years on from the initial MMSD process, the ICMM reaffirmed its work as aiming to ensure 'sustainable development fundamentals [are] embedded in company culture ... necessitating a shift from a cost culture to a value culture' (Buxton 2012, 15). This focus on value culture is a central component of the ICMM's interest to ingrain ethical business norms and practices in its member companies. As former BHP Billiton Vice President Holly Lindsay notes, at the commencement of the MMSD:

> The industry's licence to operate was very much under threat. There was an agreement to have [the MMSD] report written. And I think there was a widespread recognition of a whole range of difficult issues from both societies and the companies, [including] environment and communities. And the industry, generally, has been – as has government and the international agencies – has been working to try to address some of these issues since. And, you know, for a large part, I think performance has improved, but there are still lots of legacy issues.

One reason for legacy issues may be that the *Sustainable Development Framework* operates at the level of universal principles. It provides member companies only the vaguest advice as to what ethics might mean, deferring instead to member companies' individual principles and codes. The notes to Principle 1, for example, read (in part), 'Develop and implement company statements of ethical business principles, and practices that management is committed to enforcing'. The ICMM has taken measures to flesh out the *Framework* through the creation of position statements providing greater detail against particular principles and through the generation of practical guides, toolkits and research reports. Yet a definition of ethics remains fuzzy. While the ICMM aligns business ethics closely with corporate governance, its presentation of the subject is focused on traditional risk management, with other framework principles referring to human rights, employee and community relations and good environmental practice. While this is not to suggest that the combined principles cannot be used to guide mining sector ethics, it is a notable siloing of issues. It can be inferred that the ICMM is adopting a conservative and traditional approach to business ethics, wherein they are limited primarily to anti-corruption, legal compliance and ethical business decision-making. These business ethics concerns are imperative, but as noted earlier, the industry must also direct ethical attention to matters related to human rights and environment.

Human rights

The United Nations leads guidance for ethical consideration of human rights, especially through its UNGC initiative which defines universal principles for responsible corporate activity. Over 13,000 organizations now subscribe to the UNGC, with 189 mining companies represented among these. Since its July 2000 launch, the UNGC has set an important touchstone for business' attention to ethical and human rights issues. In an emotive speech introducing the idea of the Compact at the 1999 Davos World Economic Forum, then-Secretary General Kofi Annan challenged business leaders to, 'initiate a global compact of shared values and principles, which will give a human face to the global market' (United Nations Global Compact 2005, 14). The UNGC is steeped in this idea of universal principles and a UN commitment to hypernorms. But it was also conceived as a foundational component of a new global economy; one which might forestall restrictive regimes for global trade and investment – a move away from mandatory regulation, as discussed in Chapter 2 (Patterson 2007; United Nations Global Compact 2005). It is difficult to assess, however, the extent to which the Compact has been used as a rationale to improve or stymie regulation or sanctions. What we can say confidently is that the UNGC has been highly influential in establishing broad principles while lending the credence of a respected organization to the pursuit of socially responsible behaviour. Somewhat ironically, this is also now a main point of criticism of the UNGC, with assertions that the UN's imprimatur is exploited by companies taking advantage of a lax accountability system. Critiques like these are discussed in detail later.

Between 2000 and 2008, the UN continued to develop its work on business and human rights. By 2008, the UN Human Rights Council unanimously ratified Special Representative on Business and Human Rights, John Ruggie's, 'protect, respect, remedy' framework (Ruggie 2009, 1), embodying three principles:

> The State duty to protect against human rights abuses by third parties, including business; the corporate responsibility to respect human rights; and the need for more effective access to remedies. The three principles form a complementary whole in that each supports the others in achieving sustainable progress.

The related UN Guiding Principles on Business and Human Rights were immediately welcomed by the ICMM (Adkerson and Hodge 2011) which noted:

> The Guiding Principles offer a coherent and comprehensive basis for ensuring that the distinctive yet complimentary responsibilities of States and businesses with respect to human rights can be understood and effectively discharged.

The ICMM has since progressed operationalization of the principles through guidance documents and work against a due diligence process (International Council on Mining & Metals 2012). In encouraging this process the ICMM yokes ethical business practice to risk management and governance procedures to protect universal human rights, an approach well aligned to its association of business ethics with corporate governance. This perspective is not as limited as it first appears, however, with the ICMM suggesting that adoption of a human rights lens 'ensure[s] there are no gaps in [companies'] understanding of how environmental, economic or social impacts have the capacity to infringe upon the enjoyment of rights by others' (International Council on Mining & Metals 2012).

Similarly to the UNGC, Ruggie's framework appeals to universal principles and does not engage with the traditional debate over whether it is a corporation's place within society to uphold human rights; this responsibility is taken as rote. This positioning marks a rather underrated and important moment in the relations between corporations and human rights and between corporations and the state. As the ICMM pointed out in an open letter to John Ruggie, the mining sector was in need of 'a clear and comprehensive framework that untangles the State-business human rights nexus' (Adkerson and Hodge 2011). Questions about which societal actors hold what responsibilities and who should be responsible for providing 'access to remedies' remain controversial, but the Principles provide a strong start. And, as noted above, work to operationalize them is progressing.

Recent efforts to set out universal standards for CSR also hold the potential to assist in the operationalization of business ethics. Concerns for ethical behaviour and universal respect for human rights, embodied by the UN definitions above, for instance, are echoed in the ISO Social Responsibility Standard 26000, finalized in late 2010:[1]

> Social responsibility is the responsibility of an organization for the impacts of its decisions and activities on society and the environment, through transparent and ethical behaviour that:
>
> - contributes to sustainable development, including health and the welfare of society
> - takes into account the expectations of stakeholders
> - is in compliance with applicable law and consistent with international norms of behaviour; and
> - is integrated throughout the organization and practised in its relationships

The ISO26000 definition spurs us to question how 'international norms of behaviour' may be defined and whether such a notion can ever be enforceable. Importantly, the ISO definition makes explicit the need for transparency and disclosure and incorporates ethical corporate decision-making as an important component of business' regular responsibilities. This is significant because, in many instances, CSR is defined in terms of outcomes or impacts, with limited recognition of how it operates as a process developed in board rooms and offices, filtered through corporate policies and actioned by employees at site level. Even where a 'standard' such as this is proffered, however, the extent to which it can be translated into practice remains to be seen.

Environmental ethics

Environmental concerns and ethics are long-time bedfellows. Today, much of the wider environmental ethics discussion focuses on climate change, the scope of anthropogenic environmental damage and energy choices. Nobel Laureate and former US Vice President Al Gore's Academy Award Winning *An Inconvenient Truth* illustrates the contemporary environmental ethics landscape in which environmental issues are positioned as primarily moral, not political, considerations. Here, environmental choices are entwined with risk and intergenerational deference. Before him and echoing the Pandora myth that opened this chapter, archetypal environmentalist Rachel Carson warned that 'Our heedless and destructive acts enter into the vast cycles of the earth and in time return to bring hazard to ourselves' (Griswold 2012). Henry David Thoreau's deep relationship with Walden Pond elucidated an early environmentalist virtue ethics also centred on the place of humanity in nature (Cafaro 2004). Within each of these writings, the earth becomes an entity, one which carries with it its own necessities and responsibilities for care.

Without over-sentimentalizing, it is interesting to consider the differences in perspective which shape environmental ethicists' conceptualization of the planet against those which appear most prevalent in the global mining industry. The earth is rarely explicitly invoked in terms of ethics in mining. Instead, it is most often present as a technical object, one to be controlled and exploited. Yet, in the most basic sense, it is the centrepiece for any and all ethical decision-making concerning mining. From the use of the earth, all other impacts and decisions follow.

Environmental ethics are regularly appealed to by opponents of mining, especially where it is linked to climate change via fossil fuels or to environmental hazard. In Australia, for instance, communities opposing coal mining in the Hunter Valley, NSW, have positioned their protest in what they deemed 'an ethical framework of courage in a transnationally defined catastrophic reality' (Connor et al. 2009, 501). At the Tambogrande mining project, Peru – which has been the site of major conflict and protests with one shooting death recorded – communities founded their opposition on concern about environmental hazards. This case is particularly interesting because local communities appealed primarily to environmental ethics, eschewing calls for protection of sacred lands or invocation of ILO convention 169 concerning indigenous peoples, as is common in other, similar communities (Muradian et al. 2003). Other comparable cases abound. In Argentina, for example, the ethical dilemma for miners is made even more intense, as opposition centres not only on concerns about environmental damage but also on a general stance against mining company CSR as 'green wash' (Mutti et al. 2012).

Environmental ethics in mining are prevalent in communities not only due to broad environmental risks but also due to many communities' intimate connections to local lands. Where firms are in direct contact with communities reliant on the earth for livelihoods or for whom nature holds considerable cultural and historical value, ethical concerns compound. Anthropologists Martha Macintyre and Simon Foale, for instance, write of Melanesian activists who pursue 'popular ecological resistance … that encompasses the social meanings of resources for the people who live there' (Macintyre and Foale 2004, 231). Similarly, in a major study of mining on Lihir Island, Papua New Guinea, anthropologist and Newcrest Mining Manager of Sustainable Development and Environment, Nick Bainton, writes that mining transformed lives 'literally overnight' with the unprecedented environmental changes wrought to the island necessitating substantial compensation and relocations. Here, environmental concerns intersect business decisions in a way made clear through an ethical lens. Relocations, compensation and an extraordinary spike in material wealth challenged the very *kastom* (custom) of Lihirians and undermined 'the local moral economy' (Bainton 2010). Disparities in wealth fuelled community conflicts and changed community dynamics. In short, environmental changes necessary to the mining project precipitated socio-economic shifts that carried inherent, long-term ethical dilemmas that were not well considered prior to decisions being made.

Critiques of universal principles and voluntary codes

The mining industry's approach to ethics has been conservative, historically. Focus trends towards the more traditional business ethics areas of governance and anti-corruption (Cragg et al. 1997). This positioning is largely related to equally conservative, industry-wide perspectives on business risk. This point of view is shifting, however, as more industry leaders agree that mining activities 'precipitate risks for individuals and the communities in which they operate' (Graetz and Franks 2015, 5). Recognition of the interconnections between environmental aspects of mining, business risks, social responsibilities and ethics is also beginning to emerge. In their *Advisory Panel Report* for the 2009 *Community Relationships Review*, for example, Newmont Mining explicitly recommended that an Advisory Group could assist in determining 'how best to apply ethical analyses to the mine design and implementation process' (Newmont Mining Corporation 2009). But an extension of ethics into environmental areas, industry-wide operationalization and the regular conflation of ethics with CSR remain problematic. This is partly related to the need for an enriched understanding of how ethics is applicable to environmental and human rights spheres. It also speaks to the difficulty of applying imprecise principles to highly diverse, complicated projects.

Critics of the voluntary codes outlining ethical principles for global mining often jab the weak spots of 'free riders', adverse selection (Sethi 2005; Sethi and Emelianova 2006, 229; Sethi and Schepers 2015) and, most especially, enforceability. Business Professor S. Prakash Sethi, for instance, is pointed in his opinion of the ICMM's ineffectiveness, stating, 'the notion of voluntary codes of conduct is rendered almost worthless since it lacks any assurance of credibility and accountability' (Sethi and Emelianova 2006, 235). Certain CSR scholars, discussing broader (i.e. non-sector specific) voluntary CSR codes make similar critiques, scolding corporations for adopting codes as an instrumental means of reputation-building or legitimacy-seeking (Bondy et al. 2004; Diller 1999; Van Tulder and Kolk 2001). Critics of the UNGC are similarly scathing of its lack of enforceability and the leveraging of corporate reputation this facilitates (Entine 2010).

Yet critics of the critics argue that judgements like these miss the point and opportunity of the ICMM, UNGC and similar frameworks. Countering such appraisals of the UNGC, for instance, Warwick Business School Professor Andreas Rasche writes, 'We should not criticize the Compact for something it never pretended and/or intended to be. … The initiative is by no means a "seal of approval" for participating companies' (Rasche 2010, 17–18). The important nuance here is that universal principle frameworks are often misinterpreted in such a way as to suggest that those corporations which are members are somehow 'certified' as achieving those principles. While the ICMM and UNGC both require signatory members to report on their progress against their frameworks (ICMM via Global Reporting Initiative (GRI) sustainability reporting and UNGC through 'communications on progress' which often appear as part of GRI reports), they aim to foster ongoing consideration of ethical principles within member organizations, not to judge and censure members. As Rasche writes, if stamps of approval were to become a core part of these types of universal principle models, membership would need to be based at the individual business unit, as 'dispersed value chains' create a situation in which it is impossible to 'certify' an entire multinational corporation.

There is a conundrum here when considering ethics. If global mining companies adopt the ICMM Sustainable Development Framework, UNGC or similar guidelines for primarily reputational or legitimizing purposes; or if firms are 'free riders' to the frameworks; or if the frameworks encourage adverse selection, then are voluntary codes with low enforceability an ethical solution? Moreover, to what extent does the enforceability of ethical guidelines influence their value? What is the most constructive way to critique contemporary attempts to practice business ethics in an environment in which mining companies are to be encouraged to adopt and enact ethical principles? In other words, how might we sustain mining companies' attention to business ethics while delivering the productive but sometimes harsh critique necessary for improved ethical behaviour?

These questions are extraordinarily tricky, as their answers will depend upon the individual values and capabilities of particular mining companies, distinct local operational environments, ethical orientations of corporate leadership and the capacities of site-based communities to engage in ethical debate with the local operator. The above questions also intersect with concerns about corporate-government and corporate-community boundaries and transnational governance (Chapter 6) and means of implementing the responsible mining framework (Chapter 7). They remain open but critical and we will touch on them in the remaining chapters.

Implicating the global public in mining's ethical dilemmas

The ethical dilemmas faced by the global mining industry do not end in the pit, or even at the boundaries of affected communities. Global miners must now also consider the final uses and value chains to which they are contributing, the geopolitical environment for certain minerals, and conflict or corruption in trade related to minerals mined. This is not to suggest that the ethical responsibilities of mining companies extend indefinitely but aims instead to focus attention on the extensive horizon of the mining lifecycle, and the products and industries to which mining contributes. It is also critical to acknowledge that certain of the ethical dilemmas posed by the mining sector are linked intrinsically to global demand, implicating us all, as we benefit from mined materials in our daily lives.

Developments in mineral extraction capabilities and related technological advancements illustrate the interconnection of these dilemmas well. Where technology facilitates innovations such as mining of lower grade ore, mining on a greater scale or mining of previously inaccessible minerals, ethical decisions are pertinent. To oversimplify for the sake of illustration, such industrial advancements in minerals mined aid product improvements and inventions that are then sought out by the demand-driven market. Consumer desire for products reliant upon particular minerals drives extraction, extraction levels affect commodity prices, and prices affect supply and demand. This cycle situates mining as a vital component in a far-reaching process of consumption. But it is also one as intimate as your wrist. As BBC journalist Tim Maugham notes in his investigation into rare earth minerals, our grandmothers' watches relied on mined materials but were passed down through generations (Maughan 2015). Today, we upgrade fitbits and Apple watches as each new version becomes available, fuelling a more mineral-hungry market.

But mining companies cannot bear responsibility for global consumption. Can they? Where more or new minerals are made available, it is not up to the companies to determine what should be done with them. Rather, their focus remains on how to extract them ethically and responsibly. We enter here into a murky area in which consumer demands are in tension with resource decisions and intergenerational sustainability. The global public cannot be extracted from mining's ethical dilemmas. Instead, we are implicated in them, raising questions – beyond the bounds of this chapter – about where and how ethical resource-use decisions (e.g. access/timing/purpose) are made and the extent to which societies are involved. Here, the local community is pitted against global demand, national economies and multinational corporations.

Conclusion: some guidance towards ethical decision-making

Like the knowledge released by Pandora, mining begets hazards while simultaneously providing fundamental materials. This places industry leaders in a position requiring deliberation of morally intense issues with considerable reach. These dilemmas fall primarily into business, human rights and environmental domains, each with their own principles, guidance documents and local components. The task for the mining industry, then, is to apply ethical reasoning to synthesize these (often) competing demands in order to take ethical decisions.

The ethical decision-making literature offers sound guidance here. Meta-analyses demonstrate that ethical leadership of the corporation is critical if mining companies are to make ethical determinations. Leading business ethics researchers Michael Brown and Linda Treviño (2006, 597) note that 'moral leaders' demonstrate traits of trustworthiness, honesty and integrity, showing care in professional and personal roles. But more than the obvious:

> Moral managers make ethics an explicit part of their leadership agenda by communicating an ethics and values message, by visibly and intentionally role modeling ethical behavior, and by using the reward system (rewards and discipline) to hold followers accountable for ethical conduct. Such explicit behaviour helps the ethical leader to make ethics a leadership message that gets followers' attention by standing out as socially salient against an organizational backdrop that is often ethically neutral at best.

Research also shows that moral leadership is but one necessary component of an ethical mining company. In their wide-reaching study of ethical decision-making at the organizational level, Terry Loe and co-authors found that gender, work experience and education play negligible or inconclusive roles in influencing ethical decision-making. But age does contribute and 'culture and climate have been found to be pervasive in influencing and adapting organizational ethics' (Loe et al. 2000, 187). Organizational codes of ethics play an important role in raising awareness of moral issues and providing guidance, although the most effective structure and implementation for these codes remains debatable. Rewards and sanctions, however, were found to be effective and can contribute to repeated behaviours. Peers also play a considerable role in influencing ethical decision-making, highlighting the importance of embedding ethics throughout the corporation through training, codes of conduct and enforcement (Loe et al. 2000).

In addition to ethical leadership and organizational promotion of ethical cultures and peer modelling, theoretical models of ethical decision-making provide insight into the issues themselves. In perhaps the most widely accepted and well-developed model of ethical decision-making within organizations, theorist Thomas M. Jones (1991b) explains that ethical decisions relate to all of the above. But he goes further to argue that decisions are intrinsically linked to the characteristics of the particular issue in question. In Jones' model of 'moral intensity', the intersection between universal principles and local considerations at the heart of many mining industry dilemmas is made explicit. His model echoes certain risk matrices, to argue that for every dilemma, six particular aspects must be considered: the magnitude of consequences; social consensus; probability of effect; temporal immediacy; proximity and concentration of effect. When combined with ethical leadership qualities and strong organizational culture, bolstered by appropriate systems, ethical decision-making becomes part of doing business.

Finally, this chapter cautioned against a wholly 'cost-benefit' approach to ethics in the mining industry. Over-rationalization or purely neoliberal approaches to mining's ethical dilemmas may dehumanize local communities or, as some scholars argue, deny human rights (Hanna and Vanclay 2013). But a robust understanding of the costs of unethical behaviour, deployed as a means of ensuring strong attention to ethical concerns, offers a helpful way of utilizing an approach (economic/qualitative/technical) amenable to senior management or those colleagues in positions of power whose home languages are finance or engineering. As Rachel Davis and Daniel Franks (2014, 9) explain in their landmark study of the costs of conflict in mining:

> One company had undertaken a systematic review of the potential costs of nontechnical risks connected to its various projects and identified a significant figure – a value erosion of more than $6 billion over a two-year period, representing a double-digit per centage of its annual profits – which it used to attract Board-level attention to these issues.

The costs of unethical decision-making are clear, as are the broad tenets necessary to facilitate ethical decision-making processes within the mining company. The overview in this chapter makes a helpful start to advancing the responsible mining framework. We turn, in the next chapter, to the boundaries and governance which can further guide ethical relationships and decision-making over a project's life.

Note

1 The ISO 26000 guidance standard uses the term 'social responsibility' as opposed to 'corporate social responsibility' in order to extend the standard for use by not-for-profit groups or other non-corporate entities. See 'ISO and Social Responsibility', available: www.iso.org/iso/socialresponsibility.pdf (accessed: 7 June 2010).

References

Adkerson, R. and R. A. Hodge. 2011. 'ICMM Endorsement of the "Guiding Principles on Business and Human Rights: Implementing the UN 'Protect, Respect and Remedy" Framework.'

Aguilera, R., D. Rupp, C. Williams and J. Ganapathi. 2005. 'Putting the S Back in Corporate Social Responsibility: A Multi-Level Theory of Social Change in Organizations.' *Academy of Management Review* 32(3): 836–863.

Bainton, N. 2010. *Lihir Destiny: Cultural Responses to Mining in Melanesia*. Canberra: ANU e-Press.

Bice, S. 2015. 'Corporate Social Responsibility as Institution: A Social Mechanisms Framework.' *Journal of Business Ethics*, 1–18. DOI: 10.1007/s10551-015-2791-1.

Bird, F. and M. Velasquez. 2006. 'Introduction: Part II.' *Just Business Practices in a Diverse and Developing World: Essays on International Business and Global Responsibilities*. F. Bird and M. Velasquez, eds. Houndmills: Palgrave Macmillan, 73–80.

Blowfield, M. 2004. 'Implementation Deficits of Ethical Trade Systems.' *Journal of Corporate Citizenship* 13 : 77–90.

Bondy, K., D. Matten and J. Moon. 2004. 'The Adoption of Voluntary Codes of Conduct in MNCS: A Three-Country Comparative Study.' *Business and Society Review* 109(4): 449–477.

Brown, M. E. and L. K. Treviño. 2006. 'Ethical Leadership: A Review and Future Directions.' *The Leadership Quarterly* 17(6): 595–616.

Buxton, A. 2012. *MMSD+10: Reflecting on a Decade of Mining and Sustainable Development*. London: International Institute for Environment and Development.

Cafaro, P. 2004. *Thoreau's Living Ethics: Walden and the Pursuit of Virtue*. Athens, GA: University of Georgia Press.

Carroll, A. B. 1999. 'Corporate Social Responsibility: Evolution of a Definitional Construct.' *Business and Society* 38(3): 268–295.

Connor, L., S. Freeman and N. Higginbotham. 2009. 'Not Just a Coalmine: Shifting Grounds of Community Opposition to Coal Mining in Southeastern Australia.' *Ethnos* 74(4): 490–513.

Cragg, W., A. Greenbaum and A. Wellington. 1997. *Canadian Issues in Environmental Ethics*: Ontario: Broadview Press.

Davis, R. and D. M. Franks. 2014. 'Costs of Company-Community Conflict in the Extractive Sector.' *Corporate Social Responsibility Initiative Report*. Cambridge, MA: Harvard Kennedy School.

Diller, J. 1999. 'A Social Conscience in the Global Marketplace? Labour Dimensions of Codes of Conduct, Social Labelling and Investor Initiatives.' *International Labour Review* 138(2): 99–129.

Donaldson, T. and T. Dunfee. 1999. *Ties that Bind: A Social Contracts Approach to Business Ethics*. Massachussets: Harvard Business Review Press.

Entine, J. 2010. 'UN Global Compact: Ten Years of Greenwashing?' *Ethical Corporation's Magazine and Business Intelligence Resources*.

Franks, D. 2015. *Mountain Movers: Mining, Sustainability and the Agents of Change*. London: Routledge.

Furman, F. K. 1990. 'Teaching Business Ethics: Questioning the Assumptions, Seeking New Directions.' *Journal of Business Ethics* 9(1): 31–38.

Garriga, E. and D. Mele. 2004. 'Corporate Social Responsibility Theories: Mapping the Territory.' *Journal of Business Ethics* 53(1–2): 51–71.

Gioia, D. 1992. 'Pinto Fires and Personal Ethics: A Script Analysis of Missed Opportunities.' *Journal of Business Ethics* 11(5/6): 379–389.

Graetz, G. and D. M. Franks. 2015. 'Conceptualising Social Risk and Business Risk Associated with Private Sector Development Projects.' *Journal of Risk Research*, 1–21.

Griswold, E. 2012. 'How "Silent Spring" Ignited the Environmental Movement.' *The New York Times Magazine*.

Gunningham, N., R. Kagan and D. Thornton. 2004. 'Social License and Environmental Protection: Why Businesses Go Beyond Compliance.' *Law & Social Inquiry* 29(2): 307–341.

Hanlon, G. 2008. 'Re-Thinking Corporate Social Responsibility and the Role of The Firm: On the Denial of Politics.' *The Oxford Handbook of Corporate Social Responsibility*. A. Crane, A. McWilliams, D. Matten, J. Moon and D. Siegel, eds. Oxford: Oxford University Press.

Hanna, P. and F. Vanclay. 2013. 'Human Rights, Indigenous Peoples and the Concept of Free, Prior and Informed Consent.' *Impact Assessment and Project Appraisal* 31(2): 146–157.

Hopkins, M. 2007. *Corporate Social Responsibility and International Development: Is Business The Solution?* London: EarthScan.

International Council on Mining and Metals. 2012. *Integrating Human Rights Due Diligence into Corporate Risk Management Processes*. London: International Council on Mining and Metals.

International Organization for Standardization. 2010. 'ISO 26000 Social Responsibility Standard.'

Jones, C., M. Parker and R. T. Bos. 2005. *For Business Ethics*. London: Routledge.

Jones, T. M. 1991a. 'Ethical Decision Making by Individuals in Organizations: An Issue-Contingent Model.' *The Academy of Management Review* 16(2): 366–395.

——. 1991b. 'Ethical Decision Making by Individuals in Organizations: An Issue-Contingent Model.' *The Academy of Management Review* 16(2): 366–395.

Kant, I. 1785 [1989]. *Fundamental Principles of the Metaphysic of Morals*. London: Pearson.

Karkulehto, S. and I. Leppihalme. 2014. 'Deviant Will to Knowledge: The Pandora Myth and its Feminist Revisions.' *Deviant Women: Cultural, Linguistic and Literary Approaches to Narratives of Femininity*. T. Mantymaki, M. Rodi-Risberg and A. Foka, eds. Frankfurt am Main: Peter Lang GmbH, Internationaler Verlag der Wissenschaften.

Klonoski, R. 1991. 'Foundational Considerations in the Corporate Social Responsibility Debate.' *Business Horizons* 34(4): 9–18.

Kohlberg, L. 1984. *The Psychology of Moral Development*. New York: Harper & Row.

Kuhn, T. and S. Deetz. 2008. 'Can/Should We Get Beyond Cynical Reasoning?' *The Oxford Handbook of Corporate Social Responsibility*. A. Crane, A. McWilliams, D. Matten, J. Moon and D. Siegel, eds. Oxford University Press: Oxford.

Levinas, E. 1967 [1999]. 'Beyond Dialogue.' *Alterity and Transcendence*. M. B. Smith, ed. New York: Columbia University Press.

——. 2006. *Entre Nous: On Thinking-of-the-Other*. New York: Columbia University Press.

Loe, T. W., L. Ferrell and P. Mansfield. 2000. 'A Review of Empirical Studies Assessing Ethical Decision Making in Business.' *Journal of Business Ethics* 25(3): 185–204.

Macintyre, M. and S. Foale. 2004. 'Politicized Ecology: Local Responses to Mining in Papua New Guinea.' *Oceania* 74(3): 231–251.

Maughan, T. 2015. 'The Dystopian Lake Filled by the World's Tech Lust.' *BBC*.

Muradian, R., J. Martinez-Alier and H. Correa. 2003. 'International Capital Versus Local Population: The Environmental Conflict of The Tambogrande Mining Project, Peru.' *Society & Natural Resources* 16(9): 775–792.

Mutti, D., N. Yakovleva, D. Vazquez-Brust and M. H. Di Marco. 2012. 'Corporate Social Responsibility in the Mining Industry: Perspectives from Stakeholder Groups in Argentina.' *Resources Policy* 37(2): 212–222.

Newmont Mining Corporation. 2009. *Building Effective Community Relations: Final Report of the Advisory Panel to Newmont's Community Relationships Review*. Colorado: Newmont Mining Corporation.

Owen, D. 2005. 'CSR After Enron: A Role for the Academic Accounting Profession?' *European Accounting Review* 14(2): 395–404.

Patterson, R. 2007. 'Global Trade and Technology Regimes: The South's Asymmetrical Struggle.' *Globalization and Political Ethics*. R. B. Day and J. Masciulli, eds. Leiden: Koninklijke Brill NV.

Rasche, A. 2010. 'The UN Global Compact: A Critique of its Critiques.' *The UN Global Compact: Looking Forward after 10 Years*. M. McIntosh and S. Waddock, eds. Brisbane: Griffith University, 17–19.

Rawls, J. 1971. *A Theory of Justice*. Massachusetts: Harvard University Press.

Ruggie, J. 2009. *Business and Human Rights: Towards Operationalizing the 'Protect, Respect and Remedy' Framework*. Geneva: Human Rights Council.

Sassen, S. 2007. *A Sociology of Globalization*. New York: WW Norton & Co.

Sennett, R. 2009. *The Craftsman*. New Haven: Yale University Press.

Sethi, S. P. 2005. 'The Effectiveness of Industry-Based Codes in Serving Public Interest: The Case of the International Council on Mining and Metals.' *Transnational Corporations* 14(3): 55.

Sethi, S. P. and O. Emelianova. 2006. 'A Failed Strategy of Using Voluntary Codes of Conduct by the Global Mining Industry.' *Corporate Governance: The International Journal of Business in Society* 6(3): 226–238.

Sethi, S. P. and D. H. Schepers. 2015. 'Developing a Framework for Critiquing Multi-Stakeholder Codes Of Conduct.' *Business and the Greater Good: Rethinking Business Ethics in an Age of Crisis*. K. J. Ims and L. J. T. Pedersen, eds. Cheltenham and Northampton: Edward Elgar Publishing, 200–240.

United Nations Global Compact. 2005. 'The Global Compact Environmental Principles Training Package.' United Nations Environment Programme.

Van Tulder, R. and A. Kolk. 2001. 'Multinationality and Corporate Ethics: Codes of Conduct in the Sporting Goods Industry.' *Journal of International Business Studies* 32: 267–283.

Vogel, D. 2005. *The Market for Virtue: The Potential and Limits of Corporate Social Responsibility*. Washington, DC: Brookings Institution Press.

Weber, M. 1992 [1930]. *The Protestant Ethic and The Spirit of Capitalism*. London: Routledge.

Windsor, D. 2006. 'Corporate Social Responsibility: Three Key Approaches.' *Journal of Management Studies* 43(1): 93–114.

Wood, D., J. Logsdon, P. Lewellyn and K. Davenport. 2006. *Global Business Citizenship: A Transformative Framework for Ethics and Sustainable Capitalism*. Armonk: M. E. Sharpe, Inc.

——. 2014. *Global Business Citizenship: A Transformative Framework for Ethics and Sustainable Capitalism*. London: Routledge.

Wood, G. 2003. *Living Dolls: A Magical History For the Quest for Mechanical Life*. London: Faber & Faber.

6 Appropriate boundaries and good governance

In a remote western Australian mining town, four of the community's youngest residents burbled on a rug while their mothers made cups of tea and chatted nearby. A woman leaving her health appointment paused to coo and admire the babies. Another entered the main door, the small bells hanging from the frame jingling softly, a blast of outback air pulsing through in her wake. Off the corridor, two women carefully positioned a selection of local women's art works along the wall, preparing for the health centre's annual community art show. The centre's director, Denise,[1] continued our tour, weaving in and out of a few small clinical rooms, saying g'day to the centre's nurse, before taking a seat in the women's health library, an office-sized space heaving with pamphlets and books. For Denise, this was a regular weekday afternoon. The centre buzzed with women and young children who might otherwise become isolated in a community where many wives followed their partners' lucrative mining jobs, choosing to stay at home with their children because childcare services are sparse and expensive.

High costs of living linked to the mining operation's success, combined with very limited government support, made Denise's women's health centre heavily reliant upon funding from a local mining company. As Denise explained:

> In the last four years, [the local mining company] has put money in as part of their health partnerships and now it's reached the stage with some things that if we didn't have [their funding], some programs wouldn't be functioning because in ten years, our [government] health funding hasn't increased.

For Denise's health centre, mining company investment is a catch-22; one which is being felt in numerous other communities globally. At the country level, especially in developing countries, global miners' foreign direct investment (FDI) propels economic growth viewed as the ticket to socio-economic progress. In places like the Kyrgyz Republic – where FDI in mining contributed 12 percent of GDP, 10 percent of the national budget and half the nation's exports in 2012 (*The Economist* 2013) – the economic imperative is clear but the catch is costly. Community conflict over mining compensation is heated, locals protest environmental degradation and debates over the mine's nationalization led to the recent resignation of a Prime Minister (Jamasmie 2015; Trilling 2013). Similarly, in Mongolia, Rio Tinto-led joint venture Oyu Tolgoi is anticipated to contribute 30 percent of the country's GDP. But falling copper prices combined with high level disagreements over the mine's development saw a drop in FDI from US$4 billion in 2012 to just US$276 million in 2015 (Wen 2015). Related GDP output expanded by only 3 percent in the first half of 2015, compared to 17.3 percent growth in 2012 when the mine was tracking according to plan. As in Kyrgzstan, declining economic indicators linked to high level disputes inflamed political dissatisfaction, contributing to governmental leadership turnover at the Prime Ministerial level.

At a more local level, companies' substantial investment in local infrastructure and services and commitments to social performance raise two further requirements for responsible mining: appropriate boundaries and good governance. Extending the discussion of the preceding chapter, the establishment of appropriate corporate-community-government roles and relationships raises ethical considerations requiring pragmatic corporate decisions. Governance measures, especially those targeting social and environmental impacts, can play an important part in defining priority issues and shaping those responsibilities. This chapter begins by considering the negotiation and perimeters of appropriate boundaries and closes with a discussion of how good governance – via contemporary mining regulation and initiatives – can support clarification and enactment of those boundaries.

Boundary issues

Mining companies' improved attention to social performance is a welcome movement and company-led social development programmes are often embraced uncritically. The programmes are usually managed by companies' on-ground community relations teams and are likely to involve activities including but not limited to: local sponsorships; funding or delivery of social welfare, health or education programmes; unilateral construction of civic infrastructure, including schools, hospitals, recreation centres and non-employee housing; establishment of company-controlled trusts, funds or foundations, including NGOs; and public-private partnerships or 'outsourced' community development services. But the tendency to accept company-based social development can create unintended dependencies, lead to corporate-overreach and undermine government responsibilities (Bice 2013; Harvey and Bice 2014). If today's mining companies are bound to be socially responsible, how far does that responsibility stretch? And how can companies negotiate expanded government and stakeholder expectations about their responsibilities in an appropriate manner?

These questions are particularly vital during a time of considerable economic volatility for the mining industry. At the close of 2015, BHP Billiton shares were trading at a ten year low on Australian and New York stock exchanges. Other major miners like Newmont Gold recorded a drop in stock price from $40.12 to $17.91 during the same period. Market decline is compounded by escalating environmental concerns about fossil fuels and progressive government commitments to reduced consumption and emissions. Ahead of the November 2015 Paris climate change talks, China – the world's largest emitter – announced plans to reduce its emissions per unit of GDP by 60 to 65 percent by 2030, relative to 2005 levels (Briggs 2015). Also in advance of the Paris talks, 34 wealthy countries agreed to the OECD's plan to phase out export credit finance for coal-fired power stations, prefiguring hefty shifts toward renewable energy sources worldwide (Morton 2015). Coal heartlands, especially in the United States and Australia, are struggling through declining coal prices, shifting energy markets, consequent unemployment, and mergers and acquisitions (Krauss 2015).

In communities where mining companies make considerable contributions to infrastructure and services – especially those which would otherwise be the responsibility of government – macro-economic market volatility can have very localized consequences. For Denise's health centre – which had become reliant upon funding that is largely ad hoc and not guaranteed to be renewed – she worried for the service's viability relative to the market performance of the company. As she explained:

> [The local mining company] provides a lot of money in here now. There's no such thing as free cash but our sustainability… now, since the beginning of this year with the downturn of the mining boom, you know, I was looking at the future of the centre and a three-year contract had come up. Are we going to be able to do this if that money is gone? But that money was there and in fact, it was increased and continued. But our long-term sustainability is such an issue.

Denise's further concerns about the extent of the mining company's influence over her centre's strategic directions and programme choices echoed those raised by other members of her community in similar leadership roles (Bice 2013). Boundary creep in the form of companies' programme funding, direct involvement in social development programs or even through decisions about which social programmes to support, highlights broad ethical questions about the appropriate roles of mining companies in communities. This is not to say that companies should not prioritize social responsibilities, but instead introduces the need for a more sophisticated consideration of corporate-community-government relations. As one Africa-based senior mining company executive interviewed for this book said:

> I think there is a great temptation or a tendency to overreach in terms of the things that the company will do. … It's clear that's not a good model. Everything becomes the company's responsibility. So, if the price of lettuce is too high, that's an industrial relations issue. … I think it's a constant struggle to make sure that we find the right balance.

Research also suggests that where companies use local development as a means of short-term community pacification, communities lose out on meaningful opportunities to contribute to longer-term decisions about local mining development and its impacts (Kemp et al. 2015). This style of engagement – even where it is based less on conciliation and more on misguided good intentions – often places the company in an unenviable position of provider for the community. Such a situation can lead to conflict, especially where roles and responsibilities are unclear or out of alignment with capacity. As another senior executive interviewed for this book recounted, reflecting on experiences in West Africa:

> I once worked with a mine manager where it was the kind of community where any objective person could walk through that town, and if someone asked them, 'What does this town need?', they could probably say, 'Just about everything.' It was a subsistence, agricultural community. No sanitation. No water. No electricity. I mean, it was basic; the most basic of West African villages. ... And so, I think the mine manager there really did want to do something positive. But he just wanted to do it quickly. So, he decided that he was going to put in bore holes and water infrastructure – which, in a town with no real access to potable water, that was really positive. The problem is that he did it completely on his own, and so no coordination with local government. So, he put in infrastructure for probably four or five thousand people, which sounds fantastic. Except the community was 12,000 people. And so, immediately, everything got overused, and there was no maintenance plan. There was nothing. The bore holes had been installed, there were pumps, and everyone wanted to use it, of course. Very quickly it broke down, of course, being overused. The community came back to the mine manager and said, 'Your bore holes are broken. You need to fix them.' To which he responded, 'Those aren't mine. They're yours. You need to fix them.' [But] local government had not been involved, and there was no capacity in the local community.

On reflection, the pitfalls in the above example seem obvious. But time and again they recur in mining communities across the globe (Kemp et al. 2015). While a variety of perspectives could help to explain the persistence of situations like these, a responsible mining lens suggests that a stronger integration of evidence from leading practice impact assessments, community involvement in long-term planning through community-based agreements and corporate application of ethical decision-making processes – as detailed in the preceding chapters – would help to circumvent them. These principles have further application in negotiating the thorny terrain between supporting community and government capacity-building while avoiding paternalism.

Building capacity, limiting dependence

The ethical decision-making processes detailed in the previous chapter are vitally linked to companies' local investments and their sustainable development programming choices, even if this connection often goes unnoticed. Companies' provision of services and infrastructure is no longer a simple question of, 'Can we?'

but also 'Should we?'. Yet the 'can we' remains critical, since it is often the substantial capacity of mining companies to fill service and infrastructure gaps that leads to this dilemma in the first instance. On Lihir Island, for example, when Rio Tinto (now Newcrest) entered in 1996, the island was largely underdeveloped, necessitating the construction of roads, ports, accommodation and services to support its operations. These actions had obvious flow-on effects for Lihirians and their lifestyles, including increased use of a cash economy, changed relationships between clans and competition for market-based employment. Companies' rationales behind situations like those on Lihir, however, are in no way surprising. As social licensing expert Robert Boutilier summarizes,

> Normally what will happen is the company will be there. It will provide some roads and some of the infrastructure, which it needs [to operate] anyway. And the local government or the state or regional government will just say, 'Okay, well, we're not going to spend any of our money there because the company's there and they've got all those good paying jobs and so they don't need it as much as other regions,' which is perfectly logical. But that's what creates that problem of the boundary creep where the local people and the local government expect the company to take over government responsibilities. Meanwhile, the company is saying, 'Hey, we're only going to be here, you know, 15, 20 years; who knows how long?! And we're a resource extraction company. We don't do governance very well.'

In other words, just because corporate delivery of infrastructure and services to support the mining operation is not surprising, it does not mean it should go unquestioned. Instead, it raises a much more difficult (and admittedly unusual) question: should companies wishing to initiate projects wait until local government and community capacity reaches a state in which it is capable of supporting a mining operation? Most would argue that this presents an unrealistic scenario – it is the very capability created by the mining project that spurs the development necessary to support it. It is nevertheless interesting to contemplate, even for a moment, how different mining development and its impacts might be, were companies to initiate operations only where the appropriate capacity to support them was pre-existing.

It is crucial to question the appropriateness of company delivery of social development programmes, infrastructure and services. But it is equally important to acknowledge that lack of political will or capacity for development in certain governments can be improved through strong corporate engagement. Where roles are clear and well negotiated, appropriate company interventions may serve to build government capacity and promote socio-economic development. As Executive Director of the Institute for Human Rights and Business John Morrison details in his eloquent treatise on the 'social licence', there are also instances 'where business itself has stepped up voluntarily to show real leadership when government officials, elected or not, have demurred from doing so' (Morrison 2014). Morrison's call for corporate leadership suggests that mining companies have a considerable role to play in government capacity-building, especially in developing nations.

Morrison's concern for the role of companies in government capacity-building is playing out in many places globally. Reporting from the front line of the San Cristóbal mine in Bolivia, Robert Boutilier details a situation in which his research analysing social capital, social licence and social networks suggests that the Sumitomo owned operation is negotiating the tricky balance between capacity-building and paternalism well:

> The company is building the capacity of the communities to assume new roles as regional centres for things like health care and education and for new activities like tourism and agricultural food processing. It's helping them get the capacity, the knowledge, and the skills they need to set their own agenda because a strong community's much easier to deal with than a divided community or a weak community or a community that's afraid of change.

Where government, companies and communities focus on achieving balance, stronger social capital is cultivated, networks are strengthened and roles can be clarified. The above example is notable not only for its capturing of corporate-community-government relationships and social programme development over time (the study has been ongoing since 2009), but also for its emphasis on understanding and leveraging key networks to ensure that responsibilities are shared appropriately, roles are clear and the level (e.g. regional) of activities is appropriate to influence sustainability. Such 'bridging and bonding' through building social capital (Woolcock and Narayan 2000) also facilitates the procedural fairness which research demonstrates is necessary to establishing a 'social licence to operate' (Moffat and Zhang 2014). Consequently, corporate-led capacity-building may be effective while remaining motivated by corporate objectives. Here, we see capacity-building not for capacity's sake, but in recognition of its ability to improve company-community-government interactions and reduce conflicts and associated costs.

Establishing appropriate boundaries encourages better, deeper and more clearly defined relationships. It emphasizes reflection and adjustment, as opposed to necessarily discouraging corporate intervention in social issues. This distinction is important, as responsible mining requires action in areas beyond the traditional corporate remit. At the same time, the responsibility at its core demands careful weighing of actions to avoid creation of community dependency, absenting government responsibility and establishment of unsustainable services or infrastructure. As one South America-based senior mining executive interviewed for this book explained, the need for symmetry between pragmatic requirements and corporate drivers is also progressively influenced by high community expectations:

> I think that these are some of the main tensions that we face in the industry – it's not our business. We're not in the business of local community development. And yet, at the same time, we have to demonstrate that we bring local community development. We have to demonstrate that we are

playing a very positive role in that conversion of natural capital to human capital, to structure, all of those things. We have to play a role and oftentimes it's a tenuous balance, because expectations run very high.

The rise in expectations for mining industry behaviour and attention to concerns outside the mining pit extends beyond direct community and government relations into the realm of governance. Since at least 2002, regulation of the mining industry – through formal government legislation and voluntary initiatives – has spread considerably into social, environmental, human rights and ethical concerns. The following sections turn to these developments, detailing their potential to contribute to the good governance necessary for responsible mining and demonstrating the important role they can play in determining appropriate boundaries while also helping to define priority issues.

Transnational mining governance in the twenty-first century

The early twenty-first century distinguished a step change in mining companies' approaches to social performance. Advances in transnational mining governance – regulatory initiatives from beyond government – reflected a broader global trend in the adoption of voluntary environmental, social and governance (ESG) initiatives across many industries (Dashwood 2007; Schiavi and Solomon 2007; Vogel 2008). Such voluntary regulation comprises formal (but not mandatory) guidelines, standards, programmes or reports through which companies make public commitment and through which they must subscribe to certain norms and achieve particular benchmarks. For the mining sector, these initiatives set out sustainable development principles and best practice guidelines, diffusing social responsibility ideals across the industry largely through the social processes detailed in Chapter 2. At the turn of the century, voluntary initiatives helped global miners to redefine notions of legitimacy and construct refined criteria for the values and behaviours expected of themselves and their industry peers.

A spate of transnational governance standards and reporting frameworks fuelled advances in voluntary regulation, especially: the Mining, Minerals and Sustainable Development project (MMSD 2000–2002) which birthed the International Council on Mining and Metals (ICMM 2001); the Extractive Industries Transparency Initiative (EITI 2002); the OECD Guidelines for Multinational Enterprises (updated 2010); the United Nations Global Compact (UNGC 2010); the United Nations Guiding Principles for Business and Human Rights (Guiding Principles 2011); and the Global Reporting Initiative (GRI), including its Mining and Metals Sector Supplement (2010). Each of these initiatives has played a consequential role in encouraging responsible mining, and each could support chapter-length discussion in its own right. In the sections that follow, we will single out the ICMM, GRI and EITI as models of industry-led social performance governance, the role of reporting in facilitating accountability from voluntary initiatives, and the interplay of voluntary and mandatory regulation, respectively.

Mandatory regulation

Widely accepted voluntary initiatives coexist with a range of complex, government-led mandatory regulation that differs substantially between countries and between jurisdictions within countries. Despite differences, mandatory regulation – the statutory and legal requirements set out by governments, agencies and regulators – strongly influences companies' actions, especially in financial, environmental and traditional corporate governance areas. Although it would be impossible to canvas the breadth and depth of mandatory regulation, globally, it is worthwhile taking a moment to consider a few illustrative examples.

In Australia mandatory mining regulation is carried out through a complex regime controlled by state, territory and Commonwealth governments; a situation which arises largely due to Australia's federal structure, but which has been criticized in the past for both its duplications and inconsistencies (Sanders 2007). Broadly, Australian mining regulation focuses on management and licensing for exploration and operations, occupational health and safety (OHS) and labour standards, native title[2] and environmental and social impact assessment (ESIA).

In Canada's similarly federalist system the constitution supports contemporary legislation related to mining development, minerals access and exploitation, and communities' rights to information (Prno and Slocombe 2012). Environmental impact assessment (EIA) is enshrined in various legislation throughout Canada, with most of Northern Canada subject to some form of EIA legislation (Prno and Slocombe 2012). Canadian regulation also incorporates First Nations governments, recognizing the power and authority of the northern nation's indigenous peoples.

In South Africa, meanwhile, EIA regulations were updated in 2006 in an effort to improve their effectiveness, a measure which recent research suggests has been so far unsuccessful (Sandham et al. 2013). There, corporate attention to community issues is shaped by 1994's *Reconstruction and Development Programme*, which has resulted in many companies, especially in the platinum industry, formalizing corporate social investment (Hamann 2004). Since March 2010, South Africa is also at the forefront of mandatory 'integrated reporting' – a report which 'brings together material information about an organization's strategy, governance, performance and prospects in a way that reflects the commercial, social and environmental context within which it operates' (International Integrated Reporting Committee 2011).

The above examples only scratch the surface of what would be a very long inventory of mining-related regulation but they provide the flavour of the issue, especially in relation to ESG matters. Importantly, mandatory regulation differs not only between countries but also between areas of regulation. OHS regulation, for example, details requirements ranging from employees' personal protective equipment to measurement of total injury frequency rates. Environmental regulations cover such aspects as mine waste and tailings management, land rehabilitation and water management. The list goes on. And it represents the foundation against which transnational governance initiatives seek to build collective standards and complementary regulation.

To mandate or not to mandate?

Transnational governance initiatives address the rapidly changing situations of multinational firms in a 'world risk society' characterized by ecological, social and economic risks which pose considerable harm but which are also difficult to anticipate (Beck 1999). Consequently, these initiatives seek equipoise between addressing stakeholders' emergent concerns while limiting mandatory regulatory burden (Vogel 2008). Their widespread support within the mining sector reflects a research-identified trend in which corporations progressively separate their legal from their social obligations (Gunningham et al. 2004) in an approach to social responsibilities inclined toward risk management (Husted 2005). But the mining sector's complex operating environment raises questions about the appropriateness of voluntary regulation. It is instructive briefly to outline the debates concerning the current mix between voluntary and mandatory mining regulation before exploring the ICMM, GRI and EITI in depth.

The appeal of voluntary regulation to major mining companies is unsurprising. It allows them to acknowledge societal concerns about their industry, define social performance standards and establish monitoring bodies (such as the ICMM), all while retaining authority for monitoring and sanction within the mining industry (Sethi 2005). This arguably also 'frees' government from the task of constructing and enforcing regulation (Vogel 2008), a trend supported by countries such as Canada, Australia and the United States through 'anti-tape' initiatives (see Chapter 2 for details), often linked to arguments for supporting business productivity through regulatory efficiency. Transnational governance also facilitates application of standards and requirements across borders, potentially improving corporate behaviour, especially in developing countries where government-led requirements may be lax or capability for accountability or enforcement lacking. Echoing the boundary questions posed at the beginning of this chapter, when it comes to regulating the ESG performance of mining companies, what is an appropriate mix between voluntary and mandatory regulation? To begin answering this question, it is also helpful to consider briefly what constitutes 'good' regulation.

Deregulation vs. good regulation

There is currently a global movement for governments to pare back regulation with a view not to deregulate, but to create good regulation (Aguilera et al. 2007). This movement can be seen in the development of new regulatory tools and improved regulatory efficiencies (Haines et al. 2011), and in widespread, post-Global Financial Crisis efforts to avoid regulatory capture – wherein regulators become too close to regulated entities, compromising enforcement (OECD 2013). It also reflects shifts in attitudes about what regulation can and cannot achieve for businesses and communities. The trend further represents decreasing government control over multinational corporations that themselves have become more active political agents (Scherer and Palazzo 2011), a circumstance that has partly influenced the proliferation of transnational governance initiatives, especially in the mining industry. From a business perspective, previously accepted arguments that business gains from regulation (Stigler 1971) have largely been replaced by a neoliberal deferral to the market (Grabosky 2013) and a sense that regulation can also be counterproductive (Grabosky 1995). Alternatively, legal scholars like Christine Parker (2002) argue that it is the changing nature of the corporation and the institutionalization of responsibility which drives increased corporate self-regulation.

Still others assert that communities empowered with new media skills and the ability to hold corporations to account result in progressively stringent expectations and standards, especially in hazardous or highly impactful industries like mining (Braithwaite 2008).

The majority of transnational mining governance standards and frameworks are steeped in extensive stakeholder engagement, reflective of broader research findings indicating communities' growing expectations for corporate accountability. They also aim to be dynamically responsive to maturing stakeholder concerns in order to retain currency and legitimacy, and to secure industry-based control, free from formalized government interference (Dashwood 2007; Sethi 2005). As the MMSD Australia project noted, 'If voluntary codes are to reduce the need for government regulation, they must remain responsive to changing social conditions and stakeholder expectations' (Sheehy and Dickle 2002). Australia's Minerals Council represents this position well. In their 2006 submission to the Australian Commonwealth Government's Parliamentary Inquiry into a Sustainability Charter, the Council explicitly stated that the government would 'do better' to avoid regulation of the mining industry's social performance, where its 'effort may be better placed in encouraging other industry sectors to adopt a similar approach', as that already espoused by the MCA through activities like its own 'Enduring Value Framework' (which defines sustainable development principles), ICMM commitments, promotion of GRI reporting and support of the EITI. Its endorsement of these initiatives mirrors that of industry peak bodies and multinational miners elsewhere, and it is to these three initiatives that our attention now turns.

ICMM

Previous chapters introduced the ICMM's role in guiding the global mining industry's approach to sustainable development and its consideration of business ethics. Here, it is helpful to focus on the ICMM's influence on defining and cementing a particular approach to transnational resource governance through voluntary regulation. The ICMM's leadership in this area stems largely from the MMSD's early establishment of a widely accepted and authoritative multi-stakeholder network, involving independent, partner organizations Environmental Resources Management (ERM), the International Institute for Environment and Development (IIED) and the World Business Council for Sustainable Development

(WBCSD). These groups spearheaded a multi-stakeholder dialogue representing over 5,000 interested parties worldwide and resulting in the landmark 'Breaking new ground' report (Mining Minerals and Sustainable Development 2002). The MMSD project engaged mining industry representatives, government, NGOs, labour unions and indigenous groups to identify seven critical, non-financial issues facing the mining industry, including the need for improved governance, stakeholder engagement and recognition of indigenous people's right to free, prior and informed consent (Sheehy and Dickle 2002). By 2001, mining industry leaders coalesced to form the ICMM, committing to the sustainable development agenda articulated in the MMSD project.

The ICMM today represents 21 leading global mining companies and 33 national and regional mining associations (International Council on Mining and Metals 2014). Its governance structure is controlled whole-hog by the mining industry, a source of strength and critique (Sethi 2005). The '10 Principles for Sustainable Development' are its hallmark contribution to industry governance, with member companies required to measure and report on their performance against the Principles using the GRI sustainability reporting framework. Since 2010, companies have had the option to enhance those reports by also using the more detailed Mining and Metals Sector Supplement, developed in partnership with the GRI's multi-stakeholder network.

The Principles cover topics including ethical business practice, integration of sustainable development concerns into corporate decision-making, upholding human rights and conserving and protecting biodiversity (International Council on Mining and Metals 2010). Consequently, the ICMM describes the Sustainable Development Framework's mission as aiming 'to enhance the value creation role and long-term net contribution of the mining, minerals and metals industry and its products to people, the environment and economies' (International Council on Mining and Metals 2014, 3). Although delineation of principles like these may seem pedestrian today, at the time of the ICMM's creation they represented substantial reform and identified crucial change agents within and outside the industry (Franks 2015). The ICMM sets a critical threshold for miners' ESG performance through its strong membership base involving CEO-level commitments of many of the world's largest miners, the establishment of principles and tool kits, and progressive engagement with NGOs and intergovernmental agencies, including the UN, IFC and World Bank. Its intentions and reach have been especially influential in legitimizing a social performance agenda, championing industry-based regulation and normalizing public reporting on ESG issues, the topic to which we now turn.

To report is to regulate. Or is it? The GRI

Chapter 3 demonstrated the wealth of publicly available information on global miners' ESG performance, especially through GRI sustainability reports. It also critiqued this information and examined the extent to which it presents a thorough and trustworthy portrait of companies' activities. The analysis of Chapter 3 concluded that, while sustainability reports usually adopt a marketing perspective, shaped heavily by corporate guidelines and legal requirements, they do offer a helpful starting point for understanding the breadth and depth of the activities and values behind particular mining companies' sustainable development endeavours. The GRI sustainability reports and external assurance required by ICMM membership equate to more than a glossy document. Report production requires a full 12-month cycle, involves extensive data collection throughout all areas of a company, demands internal and external stakeholder engagement, and entreats regular attention to non-financial matters. In other words, the processes necessary to GRI report production are arguably far more valuable in advancing mining companies' self-regulation of ESG performance than the report itself.

As the GRI's Senior Strategy Advisor Maaike Fleur explained to me:

> For [GRI], a sustainability report isn't the end goal. The goal is for organizations to improve their sustainability impacts. The sustainability reporting process helps to increase awareness of those impacts and to encourage discussion with internal and external stakeholders about solutions to important challenges such as climate change, human rights issues and corruption, to name only a few.

Companies' engagement with ESG issues through the GRI reporting process is enhanced by the ICMM's requirement that they report not only against the GRI's 79 main indicators in categories including environment, society, human rights and economy, but also against the Mining and Metals Sector Supplement. The supplement is the third-most used among the GRI's 10 available sector supplements, following financial services and electric facilities. It introduces a further 11, mining-specific indicators plus mining-related commentary on the main indicators to tailor reports to mining-specific concerns.

Through its focus on issues including mine closure planning, community grievance mechanisms, artisanal mining, resettlement, indigenous rights and land use, the supplement effectively delineates expert-identified concerns. Although it is difficult to quantify the extent to which the reporting framework and related processes influence positive social performance, in my experience as a consultant on mining company sustainability reports and as a GRI-certified trainer, the reporting cycle played an important role in keeping issues on the agenda and in awareness-raising about ESG concerns throughout diverse sections of the company. GRI reports similar experiences. As Maaike Fleur describes:

> We have had a few evaluation studies, which looked at the actual impact of GRI guidelines or GRI reporting. But what we see on an empirical basis is that companies state that reporting helps them to improve their performance. Also, the ICMM has promoted the [Mining and Metals] Sector Supplement quite strongly. They have asked all their members to report with the Sector Supplement … and also to seek [external] assurance for their reports. I think that was a very brave step of ICMM to take. To my knowledge, they were the only industry-led initiative, at that time, to ask that of their members … to report using GRI.

There remains a great deal of 'in-reach' – the building of internal capacities for social performance across a mining company (Harvey and Bice 2014) – to be accomplished, however, before the considerations raised through ICMM mandates and the GRI reporting process are truly integrated across the business. Until such time, sustainability reporting will continue to play an important guiding role but will be effective as voluntary regulation only inasmuch as individual companies value the issues reported on or fear censure through public accountability. Recent changes to the GRI that focus on value chains may provide an important lever for enhancing it as a governance mechanism, however. Harvard Business School über-researchers Michael Porter and Mark Kramer (Porter and Kramer 2008) theorize that productivity in the supply chain and development of business-related clusters improve social performance. Following this, the GRI's G4 version guidelines – against which all ICMM companies must be reporting by 2015 – concentrate heavily on companies' value chains. This focus shifts the reporting process from concerns with a mining company's internally defined ESG issues toward identification and engagement of those issues seen as most material throughout its supply chain. Over time, this understanding holds the potential to address ESG impacts and benefits on a broader scale. It also presents an opportunity to convert the governance opportunities latent in the GRI reporting process, transforming it from output-directed toward outcomes-centred.

GRI reporting is also characteristic of large global miners' acquiescence to greater transparency through public disclosure. Transparency, disclosure and now 'open data' comprise another essential trait of contemporary transnational resource governance, in addition to the normative value definitions set out by the ICMM (Dashwood 2012) and the inclusion of ESG considerations into business operations as attended to through the GRI reporting process. The EITI is the most well-recognized and accepted of transparency-directed initiatives and it also represents the interplay between voluntary governance frameworks and mandatory regulation. As such, the EITI exemplifies the debate between cutting red tape and the role of government in mandating particular expectations for corporate behaviour.

The EITI

The year 2013 marked a decade since the EITI Principles were agreed at the landmark Lancaster House Conference. The Principles set out an agenda for financial transparency in the mining and extractives (e.g. oil and gas) industries that aimed to promote voluntary disclosure of resources companies' payments to governments and revenues received by those governments. In so doing, the EITI aims to reduce corruption in the hope that 'the revenue from extractive industries can transform economies, reduce poverty, and raise the living standards of entire populations in resource-rich countries' (Extractive Industries Transparency Initiative 2013, 3). The EITI represents a multi-stakeholder, 'non-state governance arrangement' (Koenig-Archibugi and Macdonald 2013) which largely responds to the 'resource curse' – the unfortunately common circumstance in which resource-rich nations suffer slower than expected or negative economic development contrary to the development potential within their resource base (Auty 1993).

Since its inception, 48 countries have implemented EITI, 27 countries are 'EITI compliant' and over US$1.3 billion of government revenue has been reported (EITI Secretariat 2014a and 2014b). The EITI is today recognized as a leading 'brand' in transparency (Scanteam 2011). Particularly in developing countries, adoption of the EITI appears to strengthen core governance or at least initiate an analysis of 'legislative, regulatory and institutional weaknesses' to allow for remediation (Scanteam 2011). Importantly, and unlike complementary reporting processes such as sustainability reporting, the EITI comprises two core components. First, companies must report what they pay to governments and governments what they have received. Second – and this is vital – EITI involves an independent, data validation process through which the figures reported are reconciled under the watch of a multi-stakeholder group. It is perhaps the multi-stakeholder model which is one of the greatest strengths of the EITI as a governance initiative. The bringing together of government, industry and civil society is a fundamental part of the process and one which the EITI's Technical and Regional Director Sam Bartlett described to me as being one of the initiative's greatest contributions, if one that often goes unrecognized:

> A new [example] is the Philippines and it's really one of the best EITI processes out there at the moment. And of course, there are some weaknesses in the EITI reporting, in terms of its comprehensiveness. But just the extent to which it's brought together government, industry and civil society. And it's got some really excellent people around the table. And what's unique is that they're not used to being in the same room, let alone developing a work plan of what they're going to do together and then reviewing the drafts together. And they're publishing something that they'll stand behind. These are companies that have been fighting with civil society for decades and now they're in the same room and they're working together.

The majority of early EITI adopters comprised developing countries, especially in Africa. This was partly due to the initiative's focus on reducing corruption and also because of its rapid rise in legitimacy (David-Barrett and Okamura 2013). Yet developing country implementation of EITI has been slow, with only two of 30 countries achieving compliance in 2010 (Natural Resource Governance Institute 2011). As of the end of 2014, 31 of 48 countries were deemed compliant with one, the Central African Republic, suspended due to noncompliance.

Very recently, developed countries are jumping onto the EITI bandwagon. At the sixth Global EITI Conference in Sydney in 2013, the UK and France made a surprise, joint announcement of their commitment to the EITI. Meanwhile, in March 2014, the US became a candidate country, while the EU has signed into law controversial requirements for country-by-country and project-by-project reporting (*Financial Times* 2012), under transparency and accounting directives, an issue we will delve into in a moment (Publish What You Pay 2013). The UK has since become an EITI candidate country, along with Ireland. As host of the most recent EITI Global Conference, Australia, rather ironically, remains a laggard in official support for EITI.

The EITI is distinctive among transnational resource governance initiatives in its focus on countries as members with mining companies as participants, as opposed to owners of the scheme. Adoption of EITI transparency norms by major developed nations is a game changer for EITI. Many of these wealthy country candidates are pursuing legislation within their home countries extending enforcement beyond national borders via their focus on where extractives firms operate. In other words, developed country legislation is essentially forcing transparency on developing countries, raising issues of national sovereignty, benign legislative colonialism and what could be called 'trickle down transparency'.

Transparent, yes. But effective?

Despite its recent growth in the number of candidate countries, EITI's effectiveness remains dubious. The EITI itself recently admitted that 'little impact [of its efforts] at a societal level can be discerned', a quandary that applies to other voluntary initiatives (Scanteam 2011). Recent studies corroborate this concern. US researcher Caitlin Corrigan (2014) finds that the EITI has no positive effect on corruption, levels of democracy, political stability or reduction of violence in member states, and shows mixed results relative to redressing the 'resource curse'. These findings echo those of the EITI itself. Its independently commissioned 2011 evaluation states that there are 'few indications that EITI programs are so far having impact on dimensions such as governance, corruption, poverty reduction or other objectives stated in EITI's *Articles of Association*' (Scanteam 2011, 3).

The above findings gel with broader discussions of transparency and accountability initiatives. Ivar Kolstad and Arne Wiig (2009) of Norway's Chr. Michelsen Institute argue that transparency alone is insufficient to kerb corruption. Writing for the OECD, Dilan Ölcer (2009) asserts that the style of revenue transparency generated by the EITI remains largely inscrutable to the general public. Related arguments were presented in a session at the 2013 EITI conference where panellists expressed concern that while necessary attention had been focused on establishing good data, this has occurred to the detriment of making that data meaningful and accessible to stakeholders. Ölcer (2009) further asserts that transparency measures falter where assumptions about a strong media watchdog or civil society are made. She argues that the EITI is being leveraged for credibility, an argument supported by Sussex University's Liz David-Barrett and the Saïd Business School's Ken Okamura (2013) who find that corrupt countries take out EITI membership with the hope of fostering reputation to increase foreign aid donations.

The EITI's recognized shortcomings raise the question of whether the initiative's stated objectives are appropriately aligned with its potential outputs and outcomes. Given that the EITI is a world-first, its initiators necessarily developed objectives based on theoretical projections about the degree and types of change it might generate. That the actual outputs and outcomes of EITI appear to differ in practice does not mean that the initiative is a failure, nor that it is not useful. At the EITI, Sam Bartlett explained to me that a key strength of the EITI is its ability to establish consistency between countries with diverse mandatory regulations while also allowing the flexibility for individual countries to address their distinct concerns:

> The starting point is: What's happening in your country? What are the challenges? Where are the information gaps? And how can the EITI add value? And there are some things you're required to do as part of the EITI process, but also you can do what you like – it's your process. And the examples where the EITI is most effective are where countries do things that may not even be mentioned in the EITI standard but which are important nationally. So, it's about providing that platform for dialogue.

Or, as the EITI's Jonas Moberg and Eddie Rich write in *Beyond Governments*, an insider's history of the EITI, 'There is no magic bullet for a problem as complex as natural resources management. It would be naïve to think that any single intervention, including the EITI, could solve the violence, mismanagement, and environmental disasters of the Niger Delta, for example. The consensus approach of the EITI standard has led to just one aspect – revenue transparency – being tackled first' (Rich and Moberg 2015, 117–118). The EITI's identified challenges, therefore, are less of a shortcoming and more of a necessary trade-off required of a multinational initiative that seeks to provide a 'one-stop-shop' while also allowing space for moulding to individual country needs. An alternative interpretation could also be that current lack of progress against stated objectives may simply be the result of the EITI's youth, with more positive results more likely given a longer timespan (Corrigan 2014).

Either way, the EITI has successfully garnered legitimacy and widespread industry and government support on a notoriously divisive topic. As Moberg and Rich write, 'In country after country, even basic revenue transparency has become the starting point for other discussions, such as whether deals are good, whether tax regimes are right, whether money is going missing, and how it is spent' (Rich and Moberg 2015, 117). Today, the EITI has efficaciously focused global attention on the critical issue of revenue transparency and influenced changes in state-based regulation, two examples of which we will now investigate.

Trickle down transparency? From EITI to legislation

In puzzling out the best regulatory mix to support responsible mining, the EITI provides a pertinent example of a voluntary, transnational governance initiative which has influenced mandatory regulation. The enshrining of resource transparency in developed countries' legislation asserts certain values and practices upon developing nations. This situation fuels national and transnational governance flows, encouraging improved financial disclosure through a process of trickle down transparency. Through the cases of the United States and European Union, we can see the channels of influence and regulatory streams that link and strengthen good resources governance.

The United States passed the Dodd-Frank Wall Street Reform and Consumer Protection Act in July 2010 in response to the failures of financial institutions in 2008's global financial crisis. The Act's Cardin-Lugar Amendment 1504 was introduced within this context of heightened corporate scrutiny. The Amendment effectively enshrines the EITI in US legislation via the Securities and Exchange Commission, requiring all NYSE-listed firms involved in the 'commercial development of oil, gas and other minerals' to report the type and total amount of payments made for each project, and the type and total amounts of payments made to each government, including taxes, royalties, licensing fees, other fees, production entitlements and any bonuses' (KPMG 2012; Publish What You Pay 2011). But the Amendment has had a rough reception and its implementation remains perilous (see Figure 6.1).

The American Petroleum Institute (the US' premier lobbying agency for the oil industry) launched a lawsuit against the SEC shortly after the Dodd-Frank Act was signed into law. The suit claimed the mandatory reporting required by the Amendment represented a breach of First Amendment rights and would make US firms less competitive against state-owned oil companies. The lawsuit, including appeals, took about three years to reach a decision. At time of writing the SEC has decided not to appeal the decision but is instead undertaking to rewrite the rules in light of the decision. In the meantime, civil society organizations, US legislators and certain investors have pressed the SEC to make speedy revisions, while related groups began petitioning Canadian legislators to pass EITI-related laws (Komnenic 2013).

Although US enforcement of transparency requirements is yet to occur, the Dodd-Frank Act pushed other governments to examine their own disclosure obligations. In 2010, the European Commission (EC) responded to Dodd-Frank by initiating action on extractives transparency. The new Accounting (and Transparency) Directive signed into law in 2013 'oblige[s] all EU listed, and large non-listed, extractive companies to publish their payments on a country and project basis' (Publish What You Pay 2013) (see Figure 6.2).

The adoption of the Accounting (and Transparency) Directive was advanced by the 2013 British Presidency of the G8. Deploying a motto of 'tax, trade, transparency', David Cameron's G8 presidency committed to an ambitious agenda. The Lough Erne Declaration secured G8 leaders' support for this agenda and set the tone for EU legislation to follow, stating, 'Extractive companies should report payments to all governments – and governments should publish income from such companies.'

Although the transparency agenda was widely supported in the G8 and the EC, the Chief Financial Officers (CFOs) of eight extractives majors voiced concerns about increased reporting requirements in a 2012 open letter to the *Financial Times*. Iterating their support for EITI, the CFOs qualified that, 'The challenge for the EU, therefore, is to ensure that the tax data published are consistent and

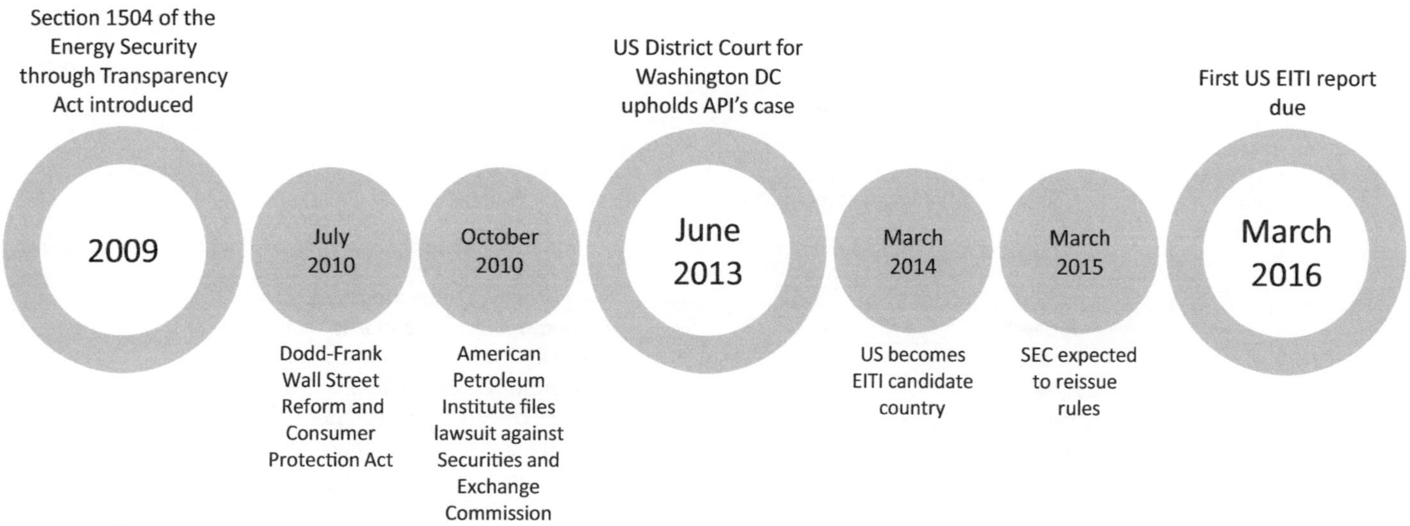

Figure 6.1 US regulation of extractives industry transparency

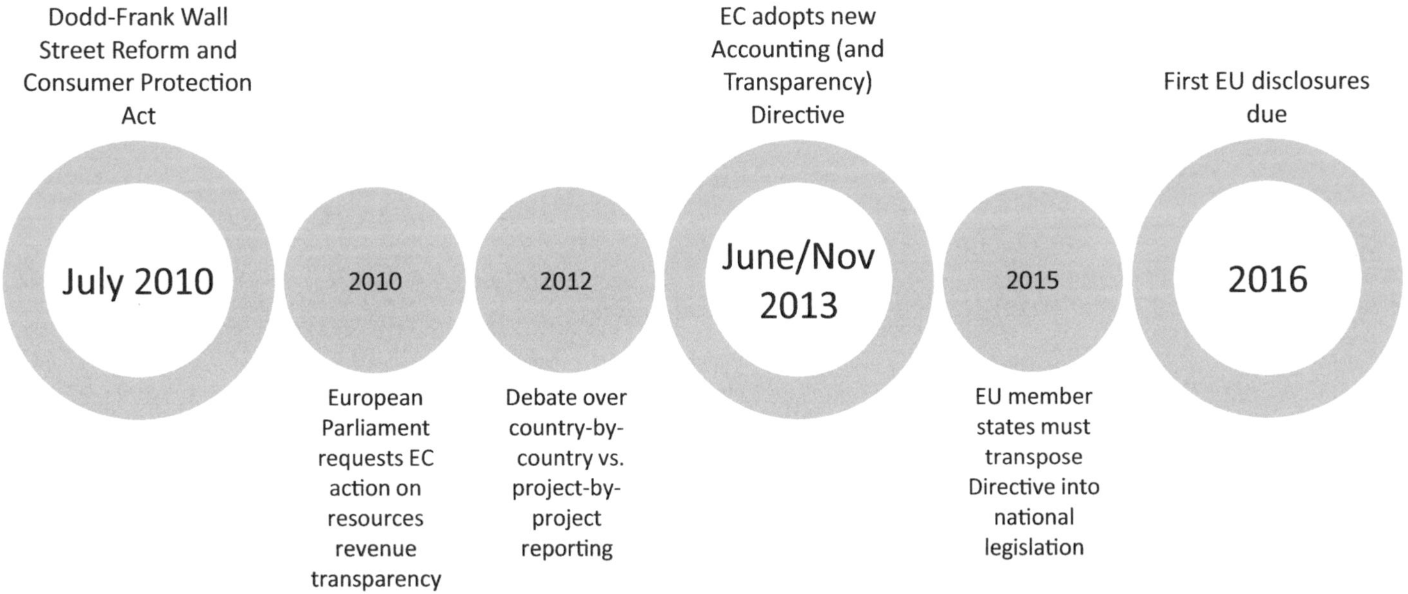

Figure 6.2 EU regulation of extractives industry transparency

comprehensive, and that they accurately reflect the flow of funds to governments' (2012). Transparency-focused civil society groups responded by arguing that: the US Dodd-Frank Act would compel project-by-project reporting anyway; that the tax structures of many corrupt African nations serve to hide monetary flows; and that while artificial payment allocations should not be generated, direct payments to particular projects should be made available publicly (2013). As national governments and extractives firms have asserted, efforts to compel country-by-country or project-by-project reporting raise questions about national sovereignty and the ability of firms to provide accurate reports where multi-jurisdictional operations place them within tax structures that are difficult to compare meaningfully to one another.

The EC legislation remains very much at a 'wait and see' moment. EU member states have until 2015 to write the Directive rules into national legislation, with disclosure information to be released beginning in 2016. So, what do these two cases tell us about the potential for trickle down transparency and how do they link to what we already know about how social drivers encourage responsible mining values and behaviours?

The same social drivers (introduced in Chapter 2) that have propelled improved social performance in the mining industry are equally instrumental to the institutionalization of transparency. Through the EITI, agreed and largely unchanged institutional principles have been disseminated through a linked group of actors (Campbell 2004), in this case the developed economy governments of the United States and EU. This process creates a 'hypernorm' (Wood et al. 2006) around transparency. At the same time, EITI values have been translated into US and EU legislation – in similar but distinct ways – allowing local, regional and state actors to make sense of a globalized framework. Peer pressure and group identity also play important roles here, with the US and EU looking to one another to drive stronger regulation of the resources sector.

Until its adoption by the US and EU, it was rare for the requirements of the EITI to be enshrined in legislation, with only Liberia and Nigeria having taken similar steps earlier (Publish What You Pay 2013). Lack of legislation could be partly related to disparate views on the nature of voluntary vs. mandatory regulation. For instance, Peter Eigen, former Transparency International head and key initiator of the EITI, argues convincingly that the distinction between 'voluntary' and 'mandatory' regulation is misleading (Eigen 2007). Countries voluntarily choose whether or not to become EITI participants, he suggests, but once signed up, compliance is mandatory. While this certainly holds in letter, the extent to which EITI may be a toothless tiger remains questionable, and others would argue that it is only through legislative enforcement that EITI compliance may be guaranteed. Indeed, in attempting to explain the EITI's lack of 'societal results', a major study suggested that this was 'largely due to lack of links with larger public sector reform processes and institutions' (Scanteam 2011).

But EITI-related legislation may also represent efforts of developed country governments to exercise greater control over activities and revenue flows occurring offshore. While the positive benefits of transparency are not to be shunned, it is also important to consider whether the benevolent veil of transparency may be used as a means of government overreach into developing states. The operationalization of transparency through legislation raises questions about the appropriateness, usefulness and potential of trickle down transparency. If – as the US and EU-proposed laws suggest – legislating transparency via a company's listing on a particular stock exchange or based on its size and headquarter location (as in the EU directive) places transparency requirements on the majority of the world's major extractives companies,[3] then we must question whether developing nations require their own transparency legislation or whether such legislation infringes upon national sovereignty. The effectiveness of the legislation and the potential for trickle down transparency remains unproven. Reflecting on regulatory mix, it also raises the question of whether mandatory transparency legislation makes the EITI redundant and also whether this matters if the desired transparency and related outcomes are achieved.

Both the US and EU cases herald the potential for transnational governance norms to generate peer and regulatory pressures to enshrine accepted norms, like transparency, in legislation. The cases equally demonstrate the sophisticated and strong debate around putting these norms into practice. Arguments raised against the US and EU legislation – harm to commercial market activity, increased costs, misalignment with tax structures and a need for improved tax collection – raise difficult questions about the best voluntary vs. mandatory mix. This tension is especially visible when we remember that the very groups which fought formalization of transparency legislation are at the same time key supporters of the EITI. Furthermore, the high level at which these debates are taking place raises important questions about who participates in policy-making discussions, especially where initiatives related to transparency are tightly linked to ideals of reducing corruption to improve citizens' lives.

The movement toward formal regulation also raises important issues of contestation. The EITI cases elevate concerns about intergovernmental channels of influence and the exercise of power and agency within these. Even the very brief examples provided here indicate a number of disparate parties involved, including: industry, civil society, developing countries, developed countries, affected communities, citizens more generally, extractives-related business (e.g. contractors), government at multiple levels, intergovernmental organizations and nongovernmental agencies. EITI's commitment to a multi-stakeholder means of operation goes some way toward addressing these concerns, but where transparency becomes legislated, this balance of power shifts toward the legislating government. The central consideration then becomes one of balancing representation and wilful participation with multi-party accountability and enforcement.

The current regulatory mix: a summary

If we are to contemplate an appropriate regulatory mix for contemporary mining, it is helpful to review the current state of play. Returning to the content analysis of sustainability reports detailed in Chapter 3, the findings revealed that voluntary initiatives dominate the contemporary regulatory landscape of the global mining industry. Across the 50 analysed reports published during the decade 2004 to 2014 by five multinational miners, the studied companies were far more likely to discuss their compliance with voluntary standards (95% of regulation discussed in all reports across all years) than mandatory regulation (5% of all regulation discussed in all reports across all years) (see Figure 6.3). Much of the voluntary regulation discussed referred to internal policies and procedures, including: risk management (especially financial) (14% of total regulation disclosures across all years) and OHS and governance policies (64% of total regulation disclosures across all years). Voluntary transnational governance initiatives made up the remainder of regulation disclosures, with the ICMM comprising seven percent of total regulation disclosures across all years, followed by the GRI (4%), ISO14001 (3%), UNGC (3%), the EITI (3%) and the UN Universal Declaration on Human Rights (2%).

The data suggests that voluntary regulation, like those detailed in the preceding sections, plays a major role in shaping companies' approaches to ESG responsibilities. Mining company representatives interviewed noted that the use of transnational governance initiatives, for which there are limited enforceable consequences, requires companies to connect their voluntary commitments to their on-ground work. Community Relations Director Michael,[4] for example, explained that voluntary frameworks provide a system of operation for the industry but need to be backed up by action at the community level: 'We need to have a system and institutionalize certain things and then, after that, it gets much more to the on-the-ground contact that you have.'

Many of the voluntary initiatives themselves dictate the importance of stakeholder engagement and the translation of espoused principles into action.

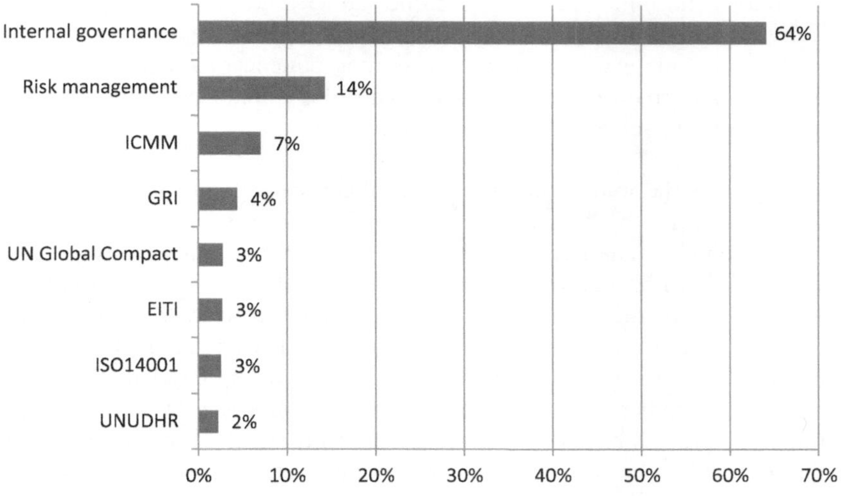

Figure 6.3 Commonly cited voluntary regulation disclosures in GRI sustainability reports studied: all years combined

The GRI, for instance, includes reporting on stakeholder identification, stakeholder engagement strategies and outcomes from those engagements as part of its 'standard disclosures' required of all reporting companies using its framework. Despite their significant focus on stakeholder engagement, however, voluntary initiatives remain subject to strong criticism from certain activist groups and NGOs (Sethi 2005). These are important critiques and they link once again to the difficult interplay of addressing local level mining impacts while managing macro-economic or country-level concerns.

Toward a sound regulatory mix

Through the preceding discussion we explored the details of three leading transnational governance initiatives, the ICMM, GRI and EITI and situated these within the current state of global mining regulation. We established that contemporary mining companies rely heavily on voluntary regulation to guide their social performance. We also know that addressing ESG issues involves the challenging task of attending to local community concerns within a global context. The beginning of this chapter contended that fuzziness between corporate-community-government roles and responsibilities flagged concerns about responsible behaviour and suggested that governance has a role to play in clarifying appropriate boundaries. Additionally, discussion from earlier chapters demonstrated that, although it is difficult to prove causal relationships, regulation is a key mechanism through which principles like responsible mining are moulded (see Chapter 2). Drawing on these lessons, is it possible to suggest what a sound regulatory framework might resemble?

The voluntary regulatory mechanisms guiding today's mining industry create a more organized and homogenized approach to companies' non-financial responsibilities and influence how companies understand and implement CSR. For Terry Heymann, Managing Director at the World Gold Council, it is the connection between micro- and macro-forces that influences industry support for transnational governance initiatives and companies' willingness to collaborate with one another to achieve the best regulatory mix:

> I think that the mining industry is definitely feeling – in a period where commodity prices have come down – it is feeling pressures around wanting to, very genuinely, mine responsibly. At the same time, it is about not getting weighted down by a mass of paperwork that actually means you're spending more time doing the documentation than actually putting the right systems in place, in terms of how you operate. And I think that's a very real and valid concern. That being said, it is important that there are appropriate systems in place and documentation in place and assurance processes in place. And so there's a very real balance there and a tension there. And that's something that everybody needs to find ways of working collaboratively to address. It's no good just kind of saying, 'We need a standard for this. We need enough guidance for this. We need a new regulation for this,' without linking in to the broader picture of what the whole environment and what the whole ecosystem around sustainability, assurance and reporting looks like.

While acknowledging the need for a suite of regulations to address the 'ecosystem' of mining's impacts, Heymann's comment also speaks to the challenge of selecting the best voluntary governance initiatives from a rapidly proliferating field. The company-based control possible through voluntary regulation must be balanced against limited propagation of standards and initiatives. Research shows that where industries are over-regulated, where limits of regulation are brought into question through 'delegalization' (Teubner 1987) via voluntary initiatives, or where companies are subject to copious regulations, the intention of such regulation may be undermined (Haines and Sutton 2003). The troubles of juridification – a process which results in the aims of governance mechanisms being lost within the minutae of administration, making compliance laborious and inefficient – detailed in Chapter 2 – remain important here. Although reliance on mandatory (and usually more centralized) regulation can assist in managing the drawbacks of too many regulations, government-led tape-cutting agendas may also be unintentionally sparking more compliance hurdles in the form of voluntary governance mechanisms.

Where balance is achieved, however, voluntary regulation may prevent further or future government intervention, create legitimation through shared practices and allow companies an operational space in which they appear regulated but through which only limited censure may occur. To be efficient and effective, voluntary initiatives must speak to one another. It is equally important that mining companies collaborate to agree on the most relevant and robust standards and encourage their shared use, while discouraging regulatory overlap between initiatives. Speaking about the World Gold Council's progress on the Conflict-Free Gold Standard, Terry Heymann articulated this challenge and also noted the role that intentional but limited 'bridging' of initiatives can play, connecting them by addressing mutual issues:

> I think it's good to be specific, and it's right to be specific, but, yes, you bridge. And you can find yourself bridging into other areas. And the challenge is how to align the standards so that they are mutually supportive and reinforcing and not duplicative. Because it doesn't benefit anybody, frankly, to have duplicative standards. So, the way that [the World Gold Council has] tried to do a lot of that in our standards is by looking to – as much as possible and as far as possible and as practical as possible – collect and base our standard off existing approaches and existing tools. And we've done that very unashamedly. Now, that won't always work … but certainly, where it does work, [it] helps to support greater consistency across different approaches. I think the most challenging thing is when you have different standards that are trying to do different things that are overlapping and duplicative. Or, indeed, when you try to get all-encompassing standards or all-encompassing approaches that somehow need to be done in addition to the sector-specific or the issue-specific standard that is in place as well. And, frankly, that's going to be an ongoing piece of work for the mining industry and it's key for much of responsible business, as to how you get that balance right.

According to Transparency International Australia's Executive Director International Greg Thompson, collective action and collective policy development by mining companies is vital to achieving optimal regulatory mix:

> I think it's a reflection of the fact that it's not just companies that are targeted, it's the whole of industry that's targeted. And people feel that we're all in this together, so let's try to address it together while we can. ... So, within the mining associations, it doesn't necessarily happen, but increasingly the standards that are set by the collective action of companies are being adopted by the member companies of those associations. ...The voluntary nature of it will be the extent to which the pressure on expectations of the wider community and investor community will have because the investors will see that these new standards are emerging that they'll hold companies to account for.

Thompson's comments articulate not only the need for collective work to govern responsible mining, but also the role in which stakeholders and investors play in holding companies to account against those standards. Former Rio Tinto Global Practice Leader Communities and Social Performance Bruce Harvey described this to me as a 'portfolio' approach to accountability. Here, mandatory regulation, stakeholders at various levels, investors and governments each play a role in ensuring that the regulatory mix ensures that companies meet diverse expectations for their performance, and that accountability flows between companies, governments and communities.

> I think, like with all things, there's got to be a recognition that a portfolio approach to holding people to account works better. Part of the portfolio is the legal and statutory requirements, and the nation state will attempt to prescribe them down to the nth degree. In my experience, if they go too far in the prescription, it will create more harm than good, because it doesn't allow enough flexibility around context-specific circumstances. So, I would rather see light regulation. I would like to then see a high degree of commitment by companies that want to be recognized as responsible to voluntary codes, and we see that. And then, when it comes to the local level, you make a common law contract with the people who are directly affected by what you're doing. ... So, I'm suggesting that this aggregates to a portfolio approach. It doesn't extend infinitely, but there are definitely layered or multi-dimensional aspects of holding a company to account, and equally holding the host governments and the host communities to account.

The potential of hybrid regulatory models incorporating voluntary and mandatory regulation to facilitate transnational resource governance while also supporting local government control is likely to grow in coming years. Recent legislative changes that mandate all or critical portions of the EITI demonstrate the likelihood that world-leading governance initiatives may have considerable and increasing influence on mandatory legislation. A sound regulatory mix, therefore, is one where collective action is used to establish and agree on frameworks used, helping to avoid duplication and consequent juridification. Through this process, a select set of broad, normative principles is established, with more specific voluntary and mandatory regulation providing mechanisms for accountability around particular issues. Such an approach facilitates a 'portfolio of accountability' that captures the views of a variety of stakeholders. It also creates channels for a flow of regulation from voluntary to mandatory, a process that facilitates introduction of innovative norms, allows for regulatory testing outside of legislation and which then improves enforcement through legislative requirement.

Conclusion

From the health centre rug to world-leading legislative chambers, appropriate boundaries and good governance play critical roles in responsible mining. In defining the final two pillars of the responsible mining framework, this chapter has demonstrated how the pillars of the preceding chapters – holistic assessment, community-based agreements and ethics – can contribute to establishing appropriate corporate-community-government boundaries and to deliberating the sound regulatory mix necessary for good governance.

Ethical reasoning, in particular, can help to define perimetres for corporate-government and corporate-community relationships that acknowledge the core values of diverse stakeholders and facilitate business viability while avoiding overreach, short-termism or building community dependence. Mining companies have a distinct opportunity to better understand their appropriate roles and responsibility in communities, through assessment consultations and community-based agreement-making processes like those presented in Chapter 4. Regulation also offers a constructive means of establishing appropriate boundaries while supporting good governance. Although voluntary regulation continues to suffer from lack of enforceability, the leading initiatives discussed in this chapter benefit from strong and widespread stakeholder engagement and accountability mechanisms. These include externally assured reports and multi-stakeholder dialogues. Examples such as the EITI also suggest that voluntary regulation may offer a proving ground for innovative regulatory solutions that can then be mandated through legislation tailored to particular countries' needs. Through these processes of responsible mining, good practice may flow from international and national regulatory levels to the doorstep of local communities. Through multi-stakeholder engagement, they may feed back from local communities to company headquarters and government.

Notes

1 This case is taken from a University of Melbourne study conducted between 2008 and 2012. The University ethics agreement guiding this study required that data be de-identified. Pseudonyms are used to protect the anonymity of respondents, in line with these requirements.

2 Although the Native Title Act is not a mining-specific legislation, the overlap of indigenous lands in remote areas and the location of ore bodies is significant, and, because of this, many native title negotiations involve mining companies.

3 Should the Dodd-Frank rules come into play, eight of the 10 largest mining companies and 29 of the 32 largest oil and gas companies will be included. The SEC reports that over half of global market capitalization in mining, oil and gas rests on US-based stock exchanges, covering over 1,100 companies.
4 This case is taken from a University of Melbourne study conducted between 2008 and 2012. The University ethics agreement guiding this study required that data be de-identified. Pseudonyms are used to protect the anonymity of respondents, in line with these requirements.

References

Aguilera, R. V., D. E. Rupp, C. A. Williams and J. Ganapathi. 2007. 'Putting the S Back in Corporate Social Responsibility: A Multi-Level Theory of Social Change in Organizations.' *Academy of Management Review* 32(3): 836–863.

Auty, R. M. 1993. *Sustaining Development in Mineral Economies: The Resource Curse Thesis*. London and New York: Routledge.

Beck, U. 1999. *World Risk Society*. Malden, MA: Polity Press.

Bice, S. 2013. 'No More Sun Shades Please: Experiences of Corporate Social Responsibility in Remote Australian Mining Communities.' *Rural Society Journal* 22(2): 138–152.

Braithwaite, J. 2008. *Regulatory Capitalism: How it Works, Ideas for Making it Work Better*. Cheltenham: Edward Elgar.

Briggs, H. 2015. 'China Climate Change Plan Unveiled.' *BBC News*.

Campbell, J. 2004. *Institutional Change and Globalization*. Princeton: Princeton University Press.

Corrigan, C. 2014. 'Breaking the Resource Curse: Transparency in the Natural Resource Sector and the Extractive Industries Transparency Initiative.' *Resources Policy* 40: 17–30.

Dashwood, H. 2007. 'Towards Sustainable Mining: The Corporate Role in the Construction of Global Standards.' *Multinational Business Review* 15(1): 47–66.

——. 2012. *The Rise of Global Corporate Social Responsibility: Mining and the Spread of Global Norms*. Cambridge: Cambridge University Press.

David-Barrett, L. and K. Okamura. 2013. 'The Transparency Paradox: Why Do Corrupt Countries Join EITI?' Oxford: Said Business School, 34.

Eigen, P. 2007. 'Fighting Corruption in a Global Economy: Transparency Initiatives in the Oil and Gas Industry.' *Houston Journal of International Law* 29(2): 337–354.

Extractive Industries Transparency Initiative. 2013. 'Articles of Association for the EITI.' Oslo: Extractive Industries Transparency Initiative, 8.

EITI Secretariat. 2014a. 'EITI Countries'. http://eiti.org/countries (accessed 3 March 2016).

——. 2014b. 'EITI Fact Sheet', https://eiti.org/files/EITI_Factsheet_EN_0.pdf (accessed 3 March 2016).

Franks, D. 2015. *Mountain Movers: Mining, Sustainability and the Agents of Change*. London: Routledge.

Financial Times. 2012. 'Project by Project Reporting Will Not Allow Citizens to "Follow the Money".'

Grabosky, P. (1995). 'Regulation by Reward: On the Use of Incentives as Regulatory Instruments.' *Law & Policy* 17(3): 257–282. DOI: 10.1111/j.1467-9930.1995.tb00150.x.

——. (2013). 'Beyond Responsive Regulation: The expanding role of non-state actors in the regulatory process.' *Regulation & Governance* 7(1): 114–123. DOI: 10.1111/j.1748-5991.2012.01147.x.

Gunningham, N., R. Kagan and D. Thornton. 2004. 'Social License and Environmental Protection: Why Businesses Go Beyond Compliance.' *Law and Social Inquiry* 29: 307–341.

Haines, F. and A. Sutton. 2003. 'The Engineer's Dilemma: A Sociological Perspective on Juridification and Regulation.' *Crime, Law and Social Change* 39: 1–22.

Hamann, R. 2004. 'Corporate Social Responsibility, Partnerships and Institutional Change: The Case of Mining Companies in South Africa.' *Natural Resources Forum* 28(4): 278–290.

Harvey, B. and S. Bice. 2014. 'Social Impact Assessment, Social Development Programmes and Social Licence to Operate: Tensions and Contradictions in Intent and Practice in the Extractive Sector.' *Impact Assessment and Project Appraisal* 32(4).

Husted, B. 2005. 'Risk Management, Real Options and Corporate Social Responsibility.' *Journal of Business Ethics* 60: 175–183.

International Council on Mining and Metals. 2010. 'Sustainable Development Framework.' www.icmm.com/our-work/sustainable-development-framework (accessed 3 March 2016).

——. 2014. 'ICMM Annual Review 2013: Strengthening Relationships with Communities.' London: ICMM.

International Integrated Reporting Committee. 2011. 'Towards Integrated Reporting: Communicating Value in the 21st Century.' Davos: IIRC.

Jamasmie, C. 2015. 'Kyrgyzstan Names New PM to Settle Issue with Centerra over Kumtor Mine.' *Mining.com*.

Kemp, D., J. R. Owen and V. Dejvongsa. 2015. 'Operational Intent and Development Impact in Mining.' *Development-Oriented Corporate Social Responsibility*. D. Jamali, C. Karam and M. Blowfield, eds. Aizlewood's Mill: Greenleaf, 49–63.

Koenig-Archibugi, M. and K. Macdonald. 2013. 'Accountability-by-Proxy in Transnational Non-State Governance.' *Governance: An International Journal of Policy, Administration, and Institutions* 26(3): 499–522.

Kolstad, I. and A. Wiig. 2009. 'Is Transparency the Key to Reducing Corruption in Resource-Rich Countries?' *World Development* 37(3): 521–532.

Komnenic, A. 2013. 'Dodd-Frank Reform on the Line: Sec Won't Appeal Ruling Against Disclosure.' *Mining.com*.

KPMG, 2012. *Financial Transparency in the Extractive Industries: Dodd-Frank Section 1504*. www.kpmg.com/US/en/IssuesAndInsights/ArticlesPublications/dodd-frank-series/Documents/financial-transparency-extractive.pdf (accessed 3 March 2016).

Krauss, C. 2015. 'Coal Miners to Struggle in an Industry Battered by Layoffs and Bankruptcy.' *The New York Times*.

Mining Minerals and Sustainable Development. 2002. 'Breaking New Ground: The Report of the MMSD.' Earthscan.

Moffat, K. and A. Zhang. 2014. 'The Paths to Social Licence to Operate: An Integrative Model Explaining Community Acceptance of Mining.' *Resources Policy* 39: 61–70.

Morrison, J. 2014. *The Social License: How to Keep Your Organization Legitimate*. London: Palgrave Macmillan.

Morton, A. 2015. 'Australia Backs Down on Coal Stand-Off.' *The Age*.

Natural Resource Governance Institute. 2011. 'The Extractive Industries Transparency Initiative (EITI).' www.resourcegovernance.org/training/resource_center/backgrounders/extractive-industries-transparency-initiative-eiti (accessed 3 March 2016).

OECD. 2013. 'OECD Economic Outlook.' Paris: OECD.

Ölcer, D. 2009. 'Extracting the Maximum from the EITI.' Paris: OECD, 52.

Parker, C. 2002. *The Open Corporation: Effective Self-Regulation and Democracy*. Cambridge and Port Melbourne: Cambridge University Press.

Porter, M. and M. Kramer. 2008. 'Strategy and Society: The Link between Competitive Advantage and Corporate Social Responsibility.' *On Competition*. Michael Porter, ed. New Haven: Harvard Business School Press, 544.

Prno, J. and D. Scott Slocombe. 2012. 'Exploring the Origins of "Social License to Operate" in the Mining Sector: Perspectives from Governance and Sustainability Theories.' *Resources Policy* 37(3): 346–357.

Publish What You Pay. 2011. 'The Cardin-Lugar Amendment (Dodd-Frank 1504)'. www.publishwhatyoupay.org/about/stock-listings/cardin-lugar-amendment-dodd-frank-1504 (accessed 30 November 2014).

——. 2013. 'Mandatory Disclosure Rules in the EU.' www.publishwhatyoupay.org/about/advocacy/mandatory-disclosure-rules-eu#1 (unavailable).

Rich, E. and J. Moberg. 2015. *Beyond Governments: Making Collective Governance Work: Lessons from the EITI*. London: Greenleaf.

Sanders, D. 2007. 'Australia's PM Seeks Unified National Mining Regulations.' *Engineering & Mining Journal* 208(1): 16.

Sandham, L. A., A. J. van Heerden, C. E. Jones, F. P. Retief and A. N. Morrison-Saunders. 2013. 'Does Enhanced Regulation Improve EIA Report Quality? Lessons from South Africa.' *Environmental Impact Assessment Review* 38(0): 155–162.

Scanteam. 2011. 'Achievements and Strategic Options: Evaluation of the Extractive Industries Transparency Initiative.' Oslo: Scanteam.

Scherer, A. G. and G. Palazzo. 2011. 'The New Political Role of Business in a Globalized World: A Review of a New Perspective on CSR and its Implications for the Firm, Governance, and Democracy.' *Journal of Management Studies* 48(4): 899–931.

Schiavi, P. and F. Solomon. 2007. 'Voluntary Initiatives in the Mining Industry: Do They Work?' *Greener Management International* (53): 27–41.

Sethi, S. P. 2005. 'The Effectiveness of Industry Based Codes in Serving Public Interest: The Case of the International Council on Mining and Metals.' *Transnational Corporations* 14(3).

Sheehy, B. and P. Dickle. 2002. 'Facing the Future: Report of the Mining Minerals and Sustainable Development Australia Project.' Melbourne: Mining Minerals and Sustainable Development, 78.

Stigler, G. J. 1971. 'The Theory of Economic Regulation' *Bell Journal of Economics and Management Science*, No. 3: 3–18

Teubner, G. 1987. 'Juridification: Concepts, Aspects, Limits, Solutions.' *Juridification of Social Spheres: A Comparative Analysis in the Areas of Labor, Corporate, Antitrust and Social Welfare Law*. Gunther Teubner, ed. Berlin: Walter de Gruyter and Co., 3–49.

The Economist. 2013. 'Mining in Kyrgyzstan: Gold in the Hills.' Asian Edition.

Trilling, D. 2013. 'Centerra Gold and Kyrgyzstan: Time for a Marriage Counsellor.' *The Globe and Mail*.

Vogel, D. 2008. 'Private Global Business Regulation.' *Annual Review of Political Science* 11(1): 261–282.

Wen, P. 2015. 'Oyu Tolgoi Is Symbol on Mongolia's Rise.' *The Sydney Morning Herald*.

Wood, D. J., J. Logsdon, P. Lewellyn and K. Davenport. 2006. *Global Business Citizenship: A Transformative Framework for Ethics and Sustainable Capitalism*. Armonk: M.E. Sharpe, Inc.

Woolcock, M. and D. Narayan. 2000. 'Social Capital: Implications for Development Theory, Research, and Policy.' *The World Bank Research Observer* 15(2): 225–249.

7 Making responsible mining happen

A theory

Our exploration of responsible mining has taken us on a journey from the assessments and community interactions of earliest exploration stages, through the trials of operational decision-making, to the negotiations of roles and responsibilities and determination of good governance. The conversation and cases throughout these discussions have remained intentionally accessible and the framework itself is pragmatically focused. Were it concealed in an obtuse intellectualism, its potential to impact the companies, communities and governments it seeks to assist would be muted. But it remains a framework strongly welded to social scientific theory. And slowly and deliberately – perhaps imperceptibly – it has been building a theory all the while.

Along the way, I suggested that the primary research agenda behind CSR drives a scholarly and practitioner obsession with establishing its business case, largely ignoring its acceptance in application. The preceding chapters are filled with examples of major companies integrating and implementing social performance into their regular business policies and practices. So, while a business case remains important, it is not all there is to understand about why and how profit-motivated corporations adopt socially responsible behaviours. For responsible mining, this point is paramount. And it is worthwhile exploring further here.

Chapter 2 established a new research agenda and model to explore how CSR has become widely integrated into industry initiatives, company policies, reporting and operational practice – even in light of a dearth of quantitative 'proof' that it contributes to business value. In pursuing a future pathway for responsible mining, Chapter 2 instead suggested that CSR is best conceptualized as an institution. As such, we can explore social drivers to understand it. A lunchroom model of CSR demonstrates that shared language, peer pressure, discipline and group identity play major roles in institutionalizing social performance within the mining industry, from peak global industry bodies to pit-side communities.

This chapter zooms out from the professionals and people, from the regulations and tribulations, to propose a middle-range theory of CSR to support future research into responsible mining. This theory presents propositions to support responsible mining's implementation. It also aims to propel future research into responsible mining. It draws together the theoretical and empirical findings of each of the preceding chapters to synthesize insights into how distinct modes of social performance develop and how they are perpetuated at various levels. The theory helps us to understand how the institutional change necessary to support responsible mining may occur. By directly connecting an enriched theorization of CSR with the responsible mining framework, the chapter provides researchers with a means of testing and understanding why responsible mining plays out as it does. These conceptual opportunities also serve to buttress the responsible mining framework with social scientific theory.

The following sections set out eight propositions for a middle-range theory of CSR (see Figure 7.1). Recalling our discussion in Chapter 2, middle-range theory is empirically focused, flexible and more interested in providing us with a means of explaining social phenomena than establishing universal rules. Here, CSR is positioned as an institution that shapes and is shaped by social contexts and

Proposition 1: CSR may be conceptualized as an institution.

Proposition 2: the mode of CSR institutionalized at any particular level of analysis may be distinctive to that institutionalized at another level.

Proposition 3: the distinct modes of CSR institutionalized at different levels of analysis are constructed through the operation and interactions of key social mechanisms with one another and with diverse social contexts.

Proposition 4: the manifestation of various modes of CSR at different levels of analysis contributes to contradictions within the institution and vice versa.

Proposition 5: identified sites of contradiction also represent sites of potential institutional change.

Proposition 6: institutional change occurs through the actions of institutional actors via the very same mechanisms which shape the institution in the first instance.

Proposition 7: institutional contradictions are not inherently negative and institutional actors may choose to leave known contradictions unaddressed because the *status quo* achieves certain industry, organizational or community aims.

Proposition 8: the next spike in institutional change in CSR will require a novel value proposition or institutional crisis.

Figure 7.1 A middle-range theory of corporate social responsibility: theoretical propositions

norms. Thinking of it as an institution allows us to consider how high level ideals intersect with on-ground social contexts and pragmatic considerations. Should responsible mining become a framework applied in practice – and I certainly hope it does, somewhere, even for a little while – the theory proposed here could be applied in future research to help us understand why it manifests in a particular way, in distinct places, at certain times. The theory proposes roles for actors within organizations and the community in influencing diverse expressions of responsible mining. It also suggests the types of situations which are most likely to support the institutional change necessary for a wholesale integration of responsible mining into core business practice. In the remainder of the chapter, we will concentrate on distilling these eight propositions, connecting pre-existing theory with the key ideas and research findings detailed throughout the book.

Proposition 1: corporate social responsibility may be conceptualized as an institution

Corporate social responsibility is not a fad. It is also more than a strategic business practice. It is an institution that has evolved over time and which, like many other institutions, takes on multifarious institutional forms. These forms are moulded by social contexts and influenced by the interactions of various, sometimes competing social mechanisms. In Chapter 2, we learned that numerous social mechanisms must interact to form an institution. This means that it is not possible to examine every mechanism which may be involved in the institutionalization of CSR (remember ideal types?). But we can apply a *bricolage* technique to identify and draw upon several core mechanisms, derived from major new institutionalist paradigms that interact to construct CSR. These include: evolution (time factors; Chapters 1–2), discourse, diffusion and translation (shared language; Chapter 3), the degree of coupling between ideals and practice (rationalized myths, decision-making; Chapters 4–5), isomorphism (peer pressure, Chapters 3–6), legitimacy (group identity, Chapters 3, 6) and regulation (discipline, Chapter 6).

As we set out theoretical propositions, below, we will reflect on the discussions of earlier chapters to link to empirical examples and ground the discussion within the framework. As shown in the first two chapters, CSR is an institution which has evolved over many years, with peaks and troughs, trends and outmoded practices. Earlier chapters clearly established that concerns for corporations' responsibilities to society are not a fly-by-night phenomenon. New institutionalists would suggest that its development represents a pattern of 'punctuated evolution', and it is worthwhile taking a moment to consider the time dimension and what this means for how we understand the development of social performance in the mining industry.

The mechanisms that shape corporate social responsibility as an institution are not only social and ideational – like the ones outlined in the lunchroom model (Chapter 2) – they are also temporal (Campbell 2004). As übersociologist, Anthony Giddens (1984) points out, institutional practices are 'those deeply embedded in time and space'. Where institutionalist studies do undertake diachronic analyses, they traditionally focus on relatively brief time periods (Campbell 2004). Too often, this results in newly studied institutions being exclaimed as revolutions. But these claims often belie the changes wrought through their slow evolution over time (Campbell 2004). By way of populist example, a researcher examining patterns of institutional change in online social networking between 2004 and 2010 might easily identify these shifts as revolutionary, centring on the explosion of Facebook and similar online tools. A researcher who demarcates a broader time horizon for social networking, incorporating its 1978 materialization in the form of IBM's Computerised Bulletin Board System (Simon 2009), however, may identify slower processes of evolutionary institutional change, within which Facebook is but one transformational development, punctuating a long history. To grossly oversimplify a concept on which an entire book could be based, new institutionalists refer to such processes as 'punctuated evolution' or punctuated equilibrium, evidenced by irregular change patterns (March and Olsen 1989). Institutions evolve when significant changes occur but the central institutional arrangements still resemble those of the past (Campbell 2004). This evolution commonly results from a few changes among institutional dimensions – which may themselves change independently and contrarily – as opposed to whole-hog institutional change (Campbell 2004).

In my opinion, this is certainly the best way to describe the establishment of CSR in the mining industry. As the development of industry frameworks discussed in Chapters 3 and 6 demonstrated, CSR in mining has been actively cultivated for decades. While the recent spike in CSR-related activities undertaken by major mining companies over the past decade – think sustainability reporting, commitment to national and international voluntary initiatives, establishment of sustainable development departments and the professionalization of community relations roles – might suggest an institutional revolution, a historical perspective counters that CSR has existed in various guises among major companies for many years. The distinction between evolution and revolution is an important one to make, as it strongly suggests that an appeal for responsible mining is not a trendy, do-gooder sentimentality. It is, instead, supported by an identifiable and theorizable institution that has evolved slowly over several decades, experiencing the peaks and troughs symptomatic of a pattern of punctuated evolutionary change.

CSR has become institutionalized largely through the convergence of various, complex social factors. As reviewed throughout the preceding chapters, these factors include: changing societal expectations for corporate behaviour; processes of globalization which support transnational operations and raise the need for international standards and guidelines; the growth of a social network society in which individuals wield greater power to voice their concerns about corporate behaviour to a global, public audience; a related confluence between traditional risk management and CSR practices; the spread of corporate policies and regulations that more directly incorporate social and environmental considerations; a rise in corporate reporting on activities that may be only indirectly related to annual financial performance; and the establishment of international initiatives and guidelines, such as the ICMM, that create in-principle behavioural benchmarks while lending the credence of respected organizations to the pursuit of socially responsible behaviour. CSR is today embedded in cultural and historical frameworks, shaped by (and shaping) cultural norms, but is not necessarily a product of strategic equilibrium or conscious design.

In asserting that CSR is an institution, the question of how the institution is influenced, operates and perpetuates is raised. We have previously explored how various social mechanisms, operating at industry, firm and community levels (here, 'levels of analysis'), interact to institutionalize CSR, making it readily identifiable, even where particular activities or policies may differ. The variety of approaches and opinions canvassed in preceding discussions elaborate core tensions and contradictions inherent in CSR. These contradictions and tensions result in three emergent institutional modes of CSR operating at the three levels of analysis: aspirational CSR, strategic CSR and tactical CSR. In these modes, we see CSR manifest as: an idealized vision of corporate-society relations; a series of policies and management practices primarily aimed at earning and maintaining a social licence to operate; and a reactive suite of programmes which are responsive to immediate local pressures, respectively.

Following this, the middle-range theory developed in this chapter suggests that the mode of CSR institutionalized at various levels of analysis is likely to be distinctive, with CSR most commonly manifesting as aspirational, strategic or tactical. We will return to these types again, later in the chapter.

Proposition 2: the mode of CSR institutionalized at any particular level of analysis may be distinctive to that institutionalized at another level

A middle-range theory of CSR suggests that various institutional modes – aspirational, strategic, tactical – are possible and that the form of CSR institutionalized in any particular context is also influenced by the level of analysis at which it operates. For example, the same social drivers that institutionalize a particular mode of CSR at the industry level may result in a very different mode of CSR being institutionalized at the community level. In all instances, the institution remains identifiable as CSR, but the degree to which certain mechanisms influence it, the specific policies and activities enacted, the gap between ideal policies and actual practice, and the regularity and extent to which these things change over time, may all be distinctive.

For example, the global principles set out through industry frameworks and transnational governance initiatives at the industry level (Chapter 6) support aspirational CSR. Aspirational CSR is steeped in an idealized vision of corporate-society relations in which the corporation is a positive social agent, achieving not only financial viability but contributing actively to economic, environmental and social sustainability of the communities in which it operates (Chapter 5).

At the corporate level, we are more likely to see a focus on strategic CSR. This mode of CSR is largely focused on those policies and activities that allow individual firms to earn and maintain a self-defined social licence to operate and to mitigate impacts and risks (Chapter 4). Strategic CSR is therefore very concerned with legitimacy gains and the management policies and approaches necessary to achieve them (Chapter 3).

Tactical CSR, meanwhile, arises primarily in response to immediate social pressures faced at the operation site level. It is often heavily focused on community relationship management, within the bounds of available resources. Circumstances particular to each site, such as local level capacity, perceptions of risk associated with particular sites, historical company-community relationships, and community capacity influence decision-making about CSR programming often resulting in reactive or ad hoc responses which struggle to address priority community needs or contribute to the sustainability to which CSR aspires. Tactical CSR may involve direct and indirect impact mitigation and building local legitimacy through successful prevention or redress of impacts (Chapter 4).

Although our ability to generalize about tactical CSR is limited by the insights possible to extract from the case studies analysed in preceding chapters, the examples discussed do suggest that tactical CSR is influenced directly by individual groups or actors who are affected by and affect CSR programming at the ground level. Tactical CSR, therefore, appears to be more responsive and reactive than the other modes of CSR. It has the potential to contribute to community benefits and reputational gains for companies. Conversely, where tactical CSR is poorly planned or implemented, it may result in ill-focused programmes, short-term (as opposed to long-term) outcomes, community rejection of or resentment towards companies' CSR initiatives, or the fostering of dependent, paternalistic relationships between mining companies and communities.

Proposition 3: the distinct modes of CSR institutionalized at different levels of analysis are constructed through the operation and interactions of key social mechanisms with one another and with diverse social contexts

At this point, we can agree that CSR is institutionalized in different modes, dependent upon the level of analysis at which it is operating. This raises the consequent question of how and why different institutional forms emerge from one particular institution? In other words, if CSR looks one way some of the time, why doesn't it look that way all of the time? It would certainly be easier for us all if we could simply pinpoint a checklist of immutable criteria, applicable to all situations and locations. 'Yes. You have a social licence to operate. Yes. This is leading edge, unquestionably good social performance.' Unfortunately, we are here building complex social theory. Them's the breaks.

Although there is no easy 'all CSR, all the time' model, the work of preceding chapters does provide us with a powerful tool for investigating the factors that influence the adoption and perpetuation of CSR, and for improving our understanding of how it operates at different levels as an institution. The framework allows us to disaggregate the complex, intertwined mechanisms – the cogs and wheels – which propel the mining industry's going concern with social performance. While facilitating examination of disparate parts, the framework also allows us to piece these mechanisms back together to construct a comprehensive understanding of CSR. A social mechanism-based framework for CSR furthers our ability to answer key questions posed throughout this book, including: what social factors influence the adoption and perpetuation of social responsibilities by mining companies? How does CSR operate in practice and what are the implications of this? If we can arm researchers with an enriched means of appreciating and explaining the processes that shape and sustain mining companies' attention to social performance this knowledge can be used to improve future engagements.

Table 7.1 summarizes the main findings about the modes of CSR institutionalized at each level of analysis, according to the social mechanisms introduced in Chapter 2. At each level of analysis, these mechanisms are operationalized differently to influence the mode of CSR we see in practice. The discussion that follows shows how these key social drivers construct distinct modes of CSR – aspirational, strategic and tactical – at various levels of analysis.

Discourse

Pervasive discourses of sustainable development, concern for society and the environment, and commitment to long-term, positive outcomes for communities at sites of operation after mine closure create a shared language of responsibility to society, chiefly through industry standards such as the ICMM, as discussed in Chapter 6. At the industry level, CSR is institutionalized through policy discourses centred on establishing and maintaining a social licence to operate and pushing individual organizations to operate beyond compliance. The cases presented throughout the book suggest that within communities where mining companies operate, discourses influence which form of CSR is institutionalized. Language can range from mutual responsibility to obligation, or something altogether different, dependent upon local context. In each instance, however, language crafts a distinctive yet identifiable form of CSR. Discursive mechanisms interact closely with mechanisms of diffusion and translation to influence how CSR spreads through different contexts.

Diffusion and translation

At all levels of analysis, ideational discourses diffuse and are translated for particular contexts. At the industry level, expectations for social performance diffuse largely through standards and industry groups. Major miners and industry bodies adopt and adapt CSR, sustainable development and community relations policies, standards, guidelines and initiatives to suit national expectations and regulatory environments (Chapter 6). At the industry level, CSR diffuses mostly in a top-down manner (evidenced by the management and governance approaches discussed in Chapter 6), and are created and revised according to input from key corporate stakeholders, including executives, senior managers and the board.

Table 7.1 Modes of CSR institutionalized at each level of analysis

Social mechanism/level of analysis *Institutional mode*	*Organizational field: industry* '*Aspirational*' CSR	*Formal organization: corporation* '*Strategic*' CSR	*Informal organization: operation site/local community* '*Tactical*' CSR
Discourse (shared language)	Cohesive, aspirational discourse generated through industry-wide initiatives and standards which draw upon credible international frameworks or guidelines Discursive focus on values, principles and models of behaviour	Idealized policy discourse centred on establishing and maintaining a social licence to operate Centralized discourse sets priorities and values	Various: in the case studies analysed here, operation site discourses range from one of mutual responsibility to one of obligation and blame
Diffusion and translation (shared language)	CSR diffuses throughout the organizational field through standards and industry groups International frameworks, such as the ICMM and UNGC, are translated into local contexts	Top-down dissemination Idealized management approaches are created and revised according to input from key corporate stakeholders, including the board Translation reflects discursive corporate values	CSR is diffused directly through company-community interactions Programmes are translated in light of local community contexts, available on-ground resources and community expectations
Decoupling (rationalized myth-making)	Policy and standards ideals create situations where gaps between industry values and practice are inevitable	Legitimacy is gained through formal policy creation and corporate headquarter messaging, but decentralized management approaches and inadequate auditing processes limit corporate headquarters' ability to ensure these policies are carried out at the operations level	The degree to which corporate policies are followed is influenced by social context at the community level, including: company size and resources, degree of professionalization of community relations staff, existing company-community relationships and accordant expectations

Social mechanism/level of analysis *Institutional mode*	*Organizational field: industry* '*Aspirational*' CSR	*Formal organization: corporation* '*Strategic*' CSR	*Informal organization: operation site/local community* '*Tactical*' CSR
Isomorphism and legitimacy (peer pressure, group identity)	Companies are encouraged to adopt centralized industry standards which draw upon the legitimacy of international frameworks and the reputation of industry associations	Corporate headquarters adapt to more cooperative, as opposed to competitive, interactions Adoption of similar policies and practices, such as the social licence, creates a normative benchmark and improves legitimacy	Companies operating within the same community adopt similar styles of CSR More homogenous approaches reinforce the legitimacy of models, even where they are decoupled from corporate headquarters' policy ideals Isomorphism may also result from competition to achieve legitimacy
Regulation (discipline)	Voluntary global regulation contributes to legitimacy and is employed as a means of 'pushing back' against the prospect of formalized regulation Involuntary regulation is based around CSR-related issues, such as social and environmental impact, OHS and labour rights	The perpetuation of traditional regulatory models, for example in relation to ESIA, limits the involuntary regulatory burden on companies but contributes to a form of CSR which is not necessarily responsive to community needs, especially over the long term	CSR programmes themselves generally operate outside of regulation This contributes to situations in which company roles and responsibilities become blurred, especially between companies and governments

At this level, individual firms translate their social performance within the context of their corporate values, missions and strategic aims, taking into consideration shareholder concerns, board objectives and the activities of peer corporations.

Chapter 6 argued that it is widely expected by governance initiatives and corporate headquarters that centrally established CSR policies and approaches will trickle down through their diverse business units or operation sites. At the community level, however, as revealed in Chapter 4, sundry factors – including a company's relationship with the local community, local expectations for corporate behaviour, residents' experiences with other industries or similarly sized companies, and the degree of agency exercised by particular community members or organizations – affect how social responsibility diffuses and is translated into community contexts. At the operation site level, at least in relation to the evidence presented throughout the book, this complex diffusion process results in social performance that is frequently more reactive than proactive, and which responds to immediate community situations.

Decoupling/loose coupling

The above processes influence the extent to which CSR, as it is practised in communities, is decoupled from the ideals and standards set out at industry and corporate headquarter levels. While variations in the degrees of decoupling or loose coupling between industry, corporate and operation site levels are unsurprising, they nevertheless contribute to the institutionalization of distinctive modes of CSR. At the industry level, the mode of CSR institutionalized is highly aspirational, as would be expected of more formal levels. The standards and initiatives noted, whether voluntary or involuntary, set out principles and expectations for values and model behaviour of mining companies in relation to their social responsibilities.

CSR is also often idealized at the corporate headquarter level. As shown in Chapter 6, value-laden forms of CSR are outlined in formal policies and through corporate headquarter messaging about commitments to sustainable development or earning and maintaining a social licence to operate (Chapter 3). At the operation site level, however, the forms of CSR institutionalized may diverge from these ideals and, in the cases examined in later chapters, are best described as only loosely coupled with the aspirational institution. In these instances, the degree to which corporate policies are followed at operation sites is influenced by specific local factors, including: company size and resources on the ground; professionalization of community relations staff; perceived levels of risk associated with particular sites; existing company-community relationships; and accordant expectations. Decoupling or loose coupling, therefore, may act in certain instances as a process of disaggregation between formal and informal organizational levels. In other instances, it may be conceived as an outcome of other mechanisms at play.

Isomorphism and legitimacy

Mechanisms of isomorphism and legitimacy also pan out differently at various levels, further influencing the appearance of distinctive modes of CSR at industry, corporate headquarter and operation site levels. At the industry level, mining companies are encouraged to adopt centralized industry standards which draw upon the legitimacy of established international frameworks, the credibility of related international organizations and the reputation of industry associations (Chapter 6).

Meanwhile, corporate headquarters are busy adapting to more cooperative, as opposed to competitive interactions. Adoption of similar policies and practices by mining companies reifies concepts, such as the social licence, and sets benchmarks for sustainability reporting and corporate action that improve legitimacy across the board (Chapter 3).

Isomorphic (normative, coercive or peer) pressures also shape the CSR-related activities carried out by companies at the operation site level. Here, the mode of CSR institutionalized may be influenced by companies operating within the same community adopting similar styles of community engagement or responding collectively to impacts (Chapter 4). These more homogenous approaches reinforce the legitimacy of the CSR models implemented, with prior, illustrative examples suggesting this is the case even where such models are known to diverge from corporate headquarters' policy ideals. This has particularly important implications for company choices about what types of CSR programmes are implemented and suggests that where multiple companies are implementing CSR programmes within the same geographic region and where community organization leaders are becoming increasingly familiar with those programmes, those leaders may begin to leverage their positions to garner better targeted programmes or greater company support. For example, local leaders, like Denise in Chapter 4, appear to be more assertive about the types of programmes they are willing to accept, while local governments are pursuing more flexible partnership arrangements. These examples highlight the potential of proactive community players to influence the forms of CSR institutionalized at the community level.

Regulation

Variations in institutional modes are also influenced by regulation, both voluntary and involuntary. For the mining industry, voluntary global regulation around social performance legitimizes financial and resource commitments to CSR and also, as suggested in Chapters 3 and 6, reduces the prospect of formalized regulation being introduced by various levels of government. At the corporate level, the perpetuation of traditional regulatory models, such as those concerning ESIA discussed in Chapter 4, also serves to limit the regulatory burden on companies and influences a form of social performance which is not necessarily responsive to community needs even though regulatory compliance may be achieved. This situation has contributed significantly to mining companies' focus on working beyond compliance to demonstrate their right to operate within communities.

At the operation site level, CSR programmes are not usually directly subject to specific regulation, barring that related to OHS, environment or bribery and corruption. In the cases presented, the lack of regulation specific to CSR may contribute to circumstances in which the roles and responsibilities of mining companies, government and key community organizations become blurred (Chapter 6). In this way, the form of CSR institutionalized at the operation site level may engender paternalistic relationships and encourage dependence upon mining companies (Chapter 4).

How might other researchers apply this framework?

At its most basic, the framework presented in Table 7.1 illustrates the importance of comparing how key social mechanisms influence the form which the institution of CSR takes at any of the three main levels of analysis. Different mechanisms or institutions may be substituted for those described here, but it is vital that the three levels of analysis outlined are maintained if meaningful social theory is to be developed (Friedland and Alford 1991). It is recommended that cases are examined at each level and that the central implications of each mechanism for the institutional mode identified are distilled. Once this disaggregate analysis is complete, the researcher is then able to compare influences and outcomes both between levels of analysis and between mechanisms. These comparisons may reveal that certain mechanisms are more influential than others at particular levels or that particular mechanisms interact more strongly with one mechanism than with others. It is very likely that such an approach will reveal contradictions within the institution as a whole, allowing the researcher to pinpoint potential sites of institutional change. These ideas are discussed further in the following propositions.

Proposition 4: the manifestation of various modes of CSR at different levels of analysis contributes to contradictions within the institution and vice versa

Let us take a moment to summarize here, as things are getting a bit intense. Proposition 3 suggested that CSR manifests in distinctive institutional modes at various levels of analysis. This occurs due to variations in how those particular mechanisms operate and interact at those levels. It then follows that contradictions may result when we look at how CSR plays out, on the whole. In other words, the intricate interactions of all those distinct mechanisms at different levels means that we're going to experience some tension. But happily, such contradictions within an institution may also act as the sites of positive institutional change (Friedland and Alford 1991). This makes it important for us to be able to identify sites of tension and contradiction in CSR. And for purposes of our middle-range theory-building, I am going to propose a two-way interaction between contradictions and institutional modes of CSR.

The multiple levels examined throughout the book reveal several crucial sites of contradiction within the institution of CSR. These sites exist *between* the different levels of analysis investigated and *within* particular levels in relation to social contexts or other institutions. Table 7.1 distils the key contradictions and tensions identified throughout our discussions and then details each throughout the following sections. In this discussion, it is also interesting to consider whether mechanisms are both sites of *creation* and contradiction? This suggestion may simply be a result of the conceptual framework applied, but it is certainly worthwhile contemplating whether social mechanisms may be conceptualized not only as the cogs and wheels which institutionalize a particular phenomenon, but also as the sites around which major tensions might revolve.

The first of many institutional tensions was detailed in Chapter 2, where CSR's widespread uptake and perpetuation among mining companies initially seemed to be more demonstrative of a fad than an institution. Yet our discussion went on to show that this paradoxical perception belies a long history of social concerns in mining. Understanding CSR as an established and evolving institution, rather than a recent, revolutionary occurrence is crucial to the assertion that it has been undergoing a slow but steady process of institutionalization in global mining. Such a stance supports later assertions that CSR policies and practices are influenced by many different social pressures and contexts, and that these contexts and their related social actors play key roles in shaping it over both time and space.

Discourse and its diffusion were also identified as a site of contradiction. In Chapter 3 the adoption of a homogenized, cohesive discourse about CSR among industry associations and through major companies' sustainability reports diverges from relationships which otherwise thrive on competitive differentiation and assertive market leadership. Related industry-wide CSR initiatives and frameworks that encourage collaboration and adoption of similar approaches to a range of issues are also largely antithetical to the highly competitive nature of the industry (Chapter 6). Similarly, the industry-level concern expressed about social issues exists in tension with a comparative lack of meaningfully measurable information on those issues and 'front-end' approaches which attempt to predict and mitigate social impacts, but which are not generally effective in responding to changing, organic social contexts (Chapter 4). Thus, while philosophical advances in the ESIA field have been made, the limited translation into mining company practice contributes to situations in which CSR rhetoric is misaligned with on-ground situations. As shown, traditional ESIA may be flawed in its use of methods which are contrary to the community engagement and long-term capacity building which the assessments encourage in principle.

At the community level, discursive contradictions appear between levels of analysis, with tensions evident in CSR discourses that are highly particular to individual operation sites and therefore necessarily divergent from more idealized and cohesive industry or corporate headquarter discourses. The diffusion of such site-specific discourses helps to reify particular CSR practices that may also diverge from corporate policies or industry-defined 'best practice' models. As a positive result, operation site-specific practices may be more tailored to particular communities. At the same time, however, such tailoring may occur without proper consideration of corporate or industry values and ideal models, resulting in CSR programmes whose outcomes are divergent from the ideal principles and intentions of such programmes (Chapters 4–5). Such discrepancies may contribute to tensions between operation sites and corporate headquarters, with decentralized management approaches and nascent monitoring and evaluation programmes resulting in little concerted action being taken to either bring site-level programmes better into line with corporate policies or to institute operation site programmes with improved, long-term outcomes for communities.

At a formal organizational level, contradictions are evident in strong corporate CSR policies whose implementation appears limited by a decentralized management approach towards policy implementation. In practice, top-down, centralized construction of CSR policy may reduce operation site employees' authority to influence policy directions and, consequently, limit their effectiveness to define appropriate measures or programme foci. This separation between corporate headquarter-designed policy, on-ground experiences and consequent management of CSR contributes to a loose coupling between organizational policy and operation site implementation, creating dilemmas for employees at formal and informal organizational levels as to whether and how CSR programmes should be carried out (Chapters 4, 6).

The implications of contradictions within CSR are perhaps nowhere more prevalent or consequential than at the operation site level. The separation between site-level programmes, corporate policies and industry-defined 'best practice' models results in CSR programmes which prize potential, often short-term legitimacy gains over values of creating communities which are sustainable in the long-term after mining. Such actions, even despite good intentions, encourage community dependence on mining companies and foster paternalistic relationships between companies and communities and, sometimes, between companies and local government (Chapter 6). It is at this site where the contradictions, tensions and dilemmas of distinctive institutional modes may result in the most significant, practical impacts, but where change may also be most possible.

Finally, contradictions are identified in the promotion of regulation which, while establishing strong expectations for socially responsible behaviour, is engendered with only limited (and sometimes no) ability in terms of enforcement. Where mandatory regulation applies to CSR-related issues, the methods associated with compliance, such as ESIA, are frequently not well-aligned with the aspirational or ideational goals or values of CSR, concentrating instead on up-front, anticipatory analyses of expected social impacts, with limited, ongoing monitoring of dynamic social situations (Chapter 4).

Although contradictions like those highlighted here might suggest that CSR is an institution so fraught with tension as to disallow meaningful reconciliation between values, policies and practice, Proposition 5 suggests instead that such sites of contradiction also represent sites of potential institutional change.

Proposition 5: identified sites of contradiction also represent sites of potential institutional change

Our discussion clearly shows that CSR is a complex institution. It is shaped by numerous mechanisms and their interactions with particular social contexts at different levels of analysis. These processes result in the development of contradictions and tensions in the institution, both between different levels of analysis and within particular levels. Contradictions, tensions, paradoxes and dilemmas are not inherently negative, however.

Tensions like those delineated throughout the preceding chapters and distilled above present opportunities for institutional change. Once particular tensions are identified, it becomes possible to consider and analyse them to determine whether action for change is necessary and to create constructive approaches which may either: reduce tensions, more actively recognize such tensions but acknowledge their existence as necessary, or perhaps change the institution more holistically in order to strengthen and perpetuate it. In the discussion below, I briefly suggest several examples where the identification of institutional contradictions might be used to advance more positive outcomes.

Tension may be reduced, for instance, by updating ESIA regulation to require more regular assessments throughout the project life and to encourage the deployment of methods which are more adaptive and responsive (Chapter 4). More regular assessment would help to address the problems which arise when ESIA is focused on a mining project's front-end and is largely anticipatory. Such a change would result in ESIA outcomes that offer a truer reflection of the current and shifting impacts communities experience throughout the life of the mining project. While many companies do now undertake more regular ESIAs or related stakeholder engagement programmes, the regularity with which these occur appears to be inconsistent and, at existing operation sites, is most likely to occur where companies are seeking licences for expansion or approval for significant changes to their operations. Adoption of more regular ESIA, using holistic assessment, could certainly reduce divergences between assessment-identified impacts, actual emergent impacts and, company responses to those issues.

In considering sites of contradiction, it is also important to acknowledge that certain tensions may be desirable. It follows that not all institutional tensions should necessarily be resolved. Industry principles for social performance and sustainable development play an important role through their establishment of aspirational standards and expectations for company behaviour. The consequent loose coupling that occurs between these standards and corporate headquarter approaches or operation site practices is to be expected. What is important here is that high principles and expectations for company behaviour are maintained, even where they are difficult to meet. The industry level is perhaps the most appropriate location for such idealized notions, and corporations and operation sites must be encouraged to continually strive to achieve the ideals set out at this level, while also acknowledging and working to address the pragmatic obstacles they face in attempting to achieve them. Balanced and transparent acknowledgement of shortcomings and challenges in meeting these ideals is, however, equally important if companies are to avoid public cynicism and accusations of greenwashing, as discussed in Chapter 3.

Finally, operation site level CSR programmes which diverge from corporate headquarter policies and implementation models suggest that operation site staff frequently find themselves in highly pressurized situations where they are forced to respond directly to community concerns and to act within very difficult environments, sometimes with limited budgets, authority or human resources. Although there was not space to present the numerous other examples of this very phenomenon I have witnessed professionally, it is worthwhile noting that I have directly observed such high pressure environments at operation sites globally, and the experiences of close colleagues confirm the situation. Moreover, previous studies which examine similar issues from ethnographic perspectives highlight these very challenges as they exist in international contexts, relative to mining companies.

In situations like those noted above, operation site staff regularly find themselves wedged uncomfortably between direct community concerns or complaints and their own professional positions and corporate models and expectations for action. Often, staff take the very human approach of reacting directly to requests or creating short-term-focused programmes that they believe will either create breathing space for their operations or stem immediate problems. These fire-fighting actions quickly become normalized among site staff, and a lack of regular, methodical corporate monitoring and evaluation to compare on-site activities against corporate policies contributes to the normalization of such practices. In many instances, as long as community-based complaints are resolved and do not make their way back up to corporate headquarters or into media, such actions are allowed to continue.

In acknowledging these tensions, actors at both the operation site and corporate levels have an opportunity to facilitate change. The gaps between strategic corporate policies and tactical operation site activities could be reduced through more concerted, regular monitoring and evaluation of social performance programmes. A more centralized management approach to programmes, or at least the diffusion of centrally approved management practices to operation sites, would help to provide operation site staff with more specific instructions as to how they are to respond to community situations and the particular types of programming which are acceptable to their companies. Equally, feedback from operation site experiences should influence corporate level policy. If it is the case that difficult on-ground situations necessitate a tactical CSR which is reactive, then perhaps corporate headquarters need formally to acknowledge this situation and to adapt their policies in such a way as to provide clearer advice on actions that are both acceptable to the corporation and will aid operation site staff to address such difficulties in more appropriate ways. Change like this could result in improved CSR programmes that generate fewer ad hoc, short-term outcomes and are better equipped to support long-term community viability.

Proposition 6: institutional change occurs through the actions of institutional actors via the very same mechanisms which shape the institution in the first instance

Institutional theory is helpful. But it retains a fairly abstracted level of focus. This means that individual and organizational actors, especially actors with agency, are sometimes marginalized in institutional theory, especially when rational-choice models are deployed (Friedland and Alford 1991). The middle-range theory of CSR proposed here incorporates 'agentic' institutional actors (Meyer 2010). These include institutional entrepreneurs (Campbell 2004; Fligstein 1997), who may draw upon the multiple and potentially competing 'institutional logics' created through contradictions and tensions in institutional form to effect institutional change (Friedland and Alford 1991). Here, actors, have a central role to play in the formation, perpetuation and transformation of institutions, and I suggest that this role may be facilitated through the very mechanisms which operate to shape the institution itself. But also let me be clear that I do not mean that individuals consciously think of or attempt to actively influence institutions through their actions. It would be the terribly odd community relations officer, concerned about the discrepancies she sees between her site's CSR programmes and those outlined by her corporate headquarter's CSR policy, who consciously decides that Tuesday will be the day on which she will finally address inherent discursive tensions to create a better integrated and less paradoxical institution (if this happens, please email me so that I can eat my hat). Instead, it is through actors' very engagement with a particular institution that they unwittingly engage mechanisms at their disposal – whether it be discourse, diffusion, modes of legitimacy or regulation – to respond to and reshape their experiences of and interactions with a particular institution.

Diffusion, for example, offers a means for actors to shift norms and expectations slowly over time. Improved corporate monitoring and evaluation of operation site programmes, like that discussed in Chapter 4, offers such a site for change. If operation site actors who are in regular contact with communities have a regular and more formalized means to provide feedback to corporate management about their experiences, then corporate management may achieve both a better understanding of the day-to-day happenings at operation sites and identify specific recommendations for improved practices. In this way, a diffusion feedback loop is created wherein strategic corporate policies trickle down to operation sites, but where operation sites are then empowered to provide meaningful advice back up the chain to influence future corporate policy-making, resource allocation and tactical decision-making. This proposition suggests that there is a micro-macro and macro-micro articulation between different organizational levels and that various modes of the one institution may influence its mode at other levels. Such two-way articulation may help to explain broader institutional characteristics, including the distinct historical eras of CSR canvassed in Chapter 2, or why CSR is so difficult to define. A diffusion feedback loop may also facilitate actors' abilities to address discrepancies between international global frameworks, industry standards, corporate headquarter policies and on-ground implementation.

Proposition 7: institutional contradictions are not inherently negative and institutional actors may choose to leave unknown contradictions unaddressed because the status quo achieves certain industry, organizational or community aims

The preceding discussion argued that contradictions represent sites of potential institutional change. But just because contradictions are identified does not necessarily mean that they will be addressed, even where such change might be beneficial for the organizations or individuals involved. This raises two important questions for consideration: first, are institutional contradictions necessarily negative, requiring redress? Secondly, why might actors support an institutional status quo, either directly or indirectly, when that status quo does not appear to be able to support optimal outcomes?

Although it is impossible to provide conclusive answers to these questions, the examples we have discussed throughout do offer a few helpful, cursory insights. In response to the first question, it seems that institutional contradictions may not only act as sites of change, they help to support complex, value-laden institutions and may also offer actors sites of agency and choice. This is discussed in relation to institutional logics, in a moment.

We can also suggest that the mode an institution takes within a particular circumstance – including whether the status quo is maintained – is strongly influenced by two central considerations: actions and values. Thus, it may be helpful to conceptualize various institutional modes as manifesting within a four-quadrant matrix defined by these two spectrums. In Figure 7.2 the X-axis describes 'models of action', with agents acting along a spectrum from 'principled' (i.e. conscientious attempts to achieve idealized practice) to 'pragmatic' (i.e. actions which respond to the circumstances at the time, based on common sense or least resistance). The Y-axis describes 'value models', ranging from 'realistic' (i.e. based on what will be acceptable within a given social context) to 'idealistic' (i.e. based on high order, aspirational values).

For example, operation sites must interact directly with communities, a position which leads them to adopt more 'pragmatic' models of action via requirements for responsiveness to immediate community needs, concerns or complaints and in which a 'realistic' value model can make it difficult for operation sites to achieve the types of ideals set out by the corporation or industry standards. As a result, operations sites tend to adopt a tactical mode of CSR which is both pragmatic and realistic. Major mining companies' corporate headquarters, meanwhile, tend to adopt strategic CSR, falling into a position that is principled but realistic. In this lower left-hand quadrant, companies set out their values and expectations for operation site behaviour in a 'principled' manner using organization-wide policies, but targets, foci and key performance indicators are based on 'realistic' values, tempered in relation to each firm's operational situations and levels of experience and success in social performance. At the macro-level, the mining industry rests in the lower right-hand quadrant and is more 'principled'/'idealistic', adopting an aspirational mode of CSR. At this level, industry associations such as the ICMM and transgovernmental agencies such as the UN, lay out idealistic universal principles and models of ideal behaviour based on those principles.

It is important to note that while it is possible to categorize each of the levels of analysis, as I have done here, these categorizations represent the most common tendencies of each level of analysis. Yet another ideal type. It is certainly also the case that some decisions regarding CSR at operation sites are highly principled and, equally, that certain decisions taken by the mining industry as a group may be based on more pragmatic considerations. In relation to future theory development, it is interesting to consider whether an institution may achieve its optimal manifestation where ideals are aligned with pragmatics, bringing idealistic values and 'real-world' situations into total alignment. The responsible mining framework supports this situation. In its ideal form, responsible mining achieves a balance between industry principles, corporate headquarter policies and community-level programs. When strongly integrated, these components engender a twenty-first century mining industry in which multi-level institutional tensions are largely reconciled.

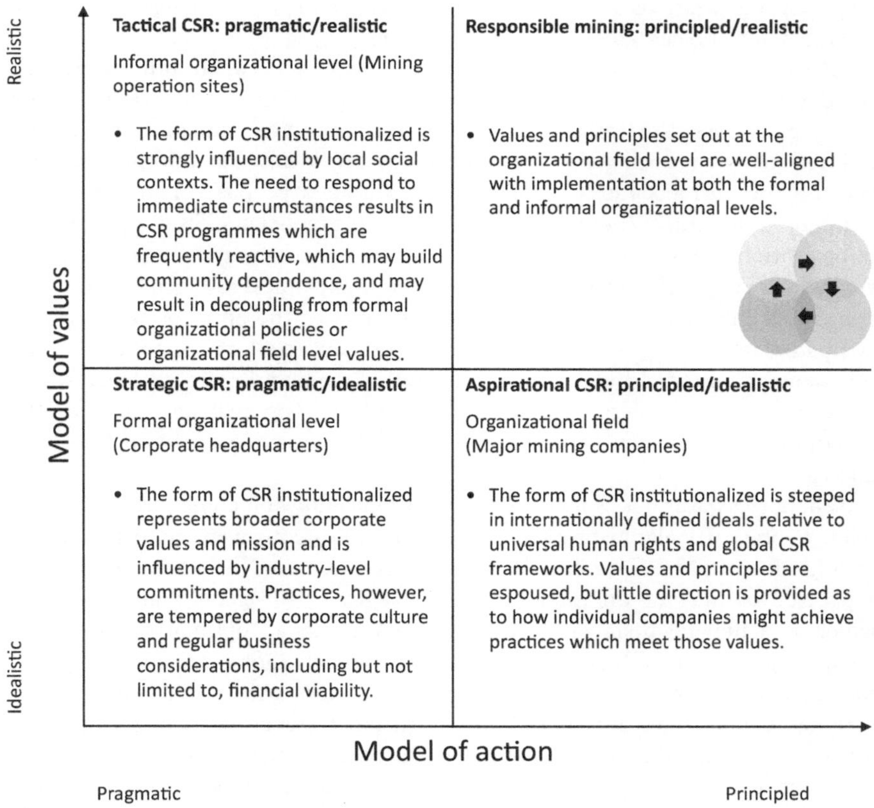

Figure 7.2 Values and actions: institutional change matrix

By visualizing institutional modes of CSR along a spectrum of pragmatic/realistic, principled/idealistic, we can begin to understand why certain contradictions are important within institutions. If, for example, attempts were made to reconcile fully the mode of CSR institutionalized at the industry level with that of the operation site, valuable principles and ideals might be lost. For example, mining companies may never be able to fully prevent or mitigate their social and environmental impacts at operation sites or to achieve situations in which they are supportive of communities without creating a degree of dependent or paternalistic relationships. But this does not mean that they should not strive to achieve these ideals. If reconciliation of tensions between ideals and on-ground situations means a loss of principles to guide action, then it is better for all parties concerned that these tensions are acknowledged but allowed to persist.

Moreover, and in response to the second question raised above, where we have actors, we also have choice. Just as actors may draw upon competing institutional logics as a means of garnering agency and creating change (Friedland and Alford 1991), such logics may also be used to support a contradiction-rife status quo. In other words, the very same multiple and competing institutional logics that empower organizations and individuals with choice to create change may also be used to make choices which deter or defer change.

Continuing with the example of the loose coupling between industry-defined standards, corporate headquarter policies and operation site practices, certain examples discussed in earlier chapters suggest that some companies are largely uninterested in reducing gaps between corporate ideals and informal practice because the current state of affairs serves a desired purpose or meets agreed, minimum standards. Where a chiefly philanthropic model of CSR is implemented in opposition to the espoused corporate approach of community engagement, for example, it may work for the community in question. Unless the social/operational context of that mine changes – for example, due to the entry of other mining companies into the local area – an operator is unlikely to alter its approach to CSR programmes. This does not, however, mean that a philanthropic approach which works at a certain operation site is necessarily the optimal approach for the corporation to adopt, but that it is sufficient. Sometimes near enough is good enough.

Where the status quo is maintained, actors may draw on competing logics that relate to one particular institution, or they may draw upon divergent logics relative to other institutions. For example, in the above scenario, the status quo may be supported by drawing on capitalist market logics that suggest that a company which is achieving financial success is operating in an optimal manner. The competing logics of CSR and capitalist institutions, therefore, are used to suggest that a change to a more advanced form of social performance at that operation site is unnecessary. Other competing, influential institutional logics might be those related to administration or to normative ideals about the autonomy of business units from the corporate parent. Following these logics, the type of monitoring and evaluation suggested to improve CSR programmes, above, may not occur regularly because of: limited operation site resources available to implement different models of CSR programmes; perceptions that more monitoring is equivalent to greater bureaucracy; the complexity of many major companies' organizational and sub-business unit structures (including complications of partial ownership); and limited corporate resources to enforce expected models. This argument is aligned with institutional change theories like Professor John Campbell's (2004), which suggest that actors may pursue or resist change based on implications for their own resources or power.

In the above scenario, numerous, competing institutional logics support a status quo. Consequently, contradictions between different levels of analysis and between distinct institutions affecting a single organization result. Although such an assertion may appear tautological, it is vital to understand institutions as organically evolving sociological phenomena which are not the result of one-way broadcasting, diffusion or mechanism-based interaction. Instead, they are better understood as cyclical; informing and informed by the key actors and social contexts in which they exist. If this were not the case, institutional change would be impossible.

Proposition 8: the next spike in institutional change in CSR will require a novel value proposition or institutional crisis

In his 2004 theory of institutional change, Professor John Campbell argues that institutional change is likely to be triggered by crises or problems arising in relation to the institution in question. In a punctuated evolutionary change pattern, such crises are often the precipitating factors behind spikes in the pattern. These crises, as Campbell (2004) defines them, may be caused by factors exogenous or endogenous to the institution, are socially constructed, may not necessarily be immediately recognized by institutional actors, and may not always result in institutional change.

For responsible mining, institutional change may be triggered not only by crises, but by the appearance of opportunities which pose new value propositions for the global industry. These opportunities may initially appear at operation sites or corporate headquarters, but they must offer a new value proposition for social performance across the entire industry in order to effect institutional change. In a roundabout way, we are returned to the CSR paradox discussed in Chapter 2. Like institutional crises, new value propositions may spur actors to seek institutional change. Historically, scholars and practitioners of CSR have been most interested in a financial business case value proposition. But as new legislation or mandatory regulation emerges; as positive reputational and more intangible gains of CSR programmes are better measured and understood; and as more people become actively involved in social performance; new, non-financial value propositions are also likely to arise. Like crises, then, positive opportunities may signal points of institutional change. Whether triggered by crises or opportunities, however, institutional actors, especially institutional entrepreneurs, are likely to play a major role in influencing institutional change (Campbell 2004).

For the mining industry, new opportunities may arise in relation to changes in international contexts and expectations. The sustainable development principles and related sustainability reporting practices championed by industry associations, for example, are likely to undergo significant changes in the next decade in light of an international push for integrated reporting – annual reports which consolidate annual financial reports with sustainability reports (Eccles et al. 2010; International Integrated Reporting Committee 2011). If, as predicted, such reporting becomes legislated by national governments or required as requisite for industry association membership, such reports will trigger a stronger focus on social performance within individual organizations; a shift which is likely to result in institutional change.

Related to this example, the GRI is also placing increased importance on local community, human rights and gender issues in its recently released G4 version of its Sustainability Reporting Guidelines and is moving towards integrated reports. Even where integrated reporting is not legislated, the strong linkages between industry standards and initiatives and GRI sustainability reporting is likely to push ICMM member companies to better integrate less tangible social considerations into their CSR programmes, a shift which could lead to institutional change at all levels.

Although as discussed in Proposition 6, it may be the case that industry and corporate headquarter actors do not wish to make changes or address institutional contradictions because the status quo is currently effective, as social performance evolves, community opinion leaders simultaneously increase their awareness of the contradictions and consequent problems that may arise from mining companies' current, on-ground approaches to community relations. As community opinion leaders become more comfortable with CSR and as they have more opportunities to see the results (or lack of results) from CSR programmes, they are more empowered to make alternative requests or even to reject particular programme options. As community opinion leaders become more assertive about shaping the types of programmes they are willing to accept and as they learn how to engage more proactively with companies' social performance programmes, we are likely to see institutional change, at both the operation site and corporate headquarter levels.

The progressive actions of mining company shareholders may also trigger institutional crises and opportunities, as more and more financial investors call upon the companies in which they place their money to demonstrate socially and environmentally responsible behaviours. The Newmont Gold example discussed in Chapter 4 demonstrates this. Companies' operation sites are also pushing for improved research on the desirability and effectiveness of their social performance activities, initiatives that may reveal new opportunities and spark changes in the forms of CSR institutionalized at the operation site level. Newmont, for example, drew upon results of its stakeholder research (Newmont Mining Corporation 2009) to shape its interactions with the Waihi community in New Zealand, which many CSR practitioners agree is one of the most advanced and well-integrated CSR programmes implemented. In brief, the programme, known locally as 'The Waihi Vision' is long-term focused, is 'owned' by the local community, is facilitated by an independent firm specializing in community engagement and, in just over a decade, has already achieved significant outcomes for the local community, including the creation of alternative revenue streams from tourism.

In Papua New Guinea, Australian mining company, Newcrest Gold, is also utilizing research and community engagement to identify opportunities to change and improve its relationships with the local community. The company currently offers the people of Lihir Island the most comprehensive benefit package ever negotiated in PNG (Bainton 2010), but it has also been recognized that the package, despite its stated focus on sustainable development, is not achieving the agreed goals it set out. In light of this, and in accordance with an agreement to regularly review the compensation and benefit package, Newcrest undertook in-depth research with community opinion leaders to determine what has worked and has not worked in relation to the package, the community's objectives for the medium- to long-term, and to construct a revised, community-centred plan to achieve these goals.[1] In so doing, the company is leveraging new opportunities for community engagement and is establishing new agreements, policies and practices.

Summarizing a middle-range theory of CSR

The middle-range theory advanced in this chapter argues that CSR is an institution shaped by a range of social mechanisms. These operate simultaneously at macro-, meso- and micro-levels to support a pattern of policies and activities enacted by corporations. The mechanisms interact in such a way as to construct a complex institution filled with internal tensions. Yet these contradictions are also the sites of potential transformation. The theory suggests that it is through the process of understanding and addressing institutional contradictions – and the mechanisms through which they operate – that the responsible mining framework can be enacted, contracting the gap between social performance principles and practice.

The eight propositions for a middle-range theory of CSR developed here assert that it may be best conceptualized as an institution that manifests in various modes, at different levels. These three distinct modes of CSR – aspirational, strategic and tactical – are constructed through the operation and interactions of key mechanisms with one another and with diverse social contexts. The interactions of diverse modes contributes to contradictions within CSR as an institution, between different levels of analysis, and vice versa.

By identifying sites of contradiction, however, we also identify sites of potential institutional change. The theory suggests that this can occur through institutional actors via the very same mechanisms that shape the institution in the first instance. Importantly, our discussion showed that institutional contradictions are not inherently negative and actors may choose to leave known contradictions unaddressed because the status quo achieves certain aims.

The middle-range theory of CSR also suggests that we require a novel value proposition or institutional crisis to spark the next institutional change for social performance in the mining industry. While some suggestions as to what this might entail have been made, it is up to future research to pinpoint the next catalyst in CSR's development.

Achieving responsible mining will require much change. And it needs to include: adoption of holistic assessment to understand impacts; company-community relationships guided by community-based agreement-making; application of ethical decision-making; the setting out of appropriate corporate-community-government roles through capacity building of communities and local governments to improve their agency and ability to engage effectively with large corporations, while discouraging companies from assuming quasi-governmental roles; and support for policies and regulations that clarify and embed good governance and accountability. At the beginning of the twenty-first century, the opportunities are there. Let's hope the institutional change required will be sparked through their proactive embrace, not the crises that would command them.

Note

1 In September 2010, I contributed to this research in a professional capacity, resulting in a community opinion leader-approved report capturing the key benefits and failings of the company's major CSR programme and suggesting community-defined ways forward.

References

Bainton, N. 2010. *The Lihir Destiny: Cultural Responses to Mining in Melanesia*. Canberra: ANU e-Press.

Campbell, J. 2004. *Institutional Change and Globalization*. Princeton: Princeton University Press.

Eccles, R., M. Krzus and D. Tapscott. 2010. *One Report: Integrated Reporting for a Sustainable Strategy*. Hoboken: John Wiley & Sons, Inc.

Fligstein, N. 1997. 'Social Skill and Institutional Theory.' *American Behavioral Scientist* 40(4): 397.

Friedland, R. and R. Alford. 1991. 'Bringing Society Back In: Symbols, Practices, and Institutional Contradictions.' *The New Institutionalism in Organizational Analysis*. P. DiMaggio and W. Powell, eds. Chicago: The University of Chicago Press, 232–263.

Giddens, A. 1984. *The Constitution of Society: Outline of the Theory of Structuration*. Cambridge: Polity.

International Integrated Reporting Committee. 2011. 'What Is Integrated Reporting?'. http://integratedreporting.org/what-the-tool-for-better-reporting/ (accessed 3 March 2016).

March, J. G. and J. P. Olsen. 1989. *Rediscovering Institutions: The Organizational Basis of Politics*. New York: Free Press.

Meyer, J. W. 2010. 'World Society, Institutional Theories, and the Actor.' *Annual Review of Sociology* 36: 1–20.

Newmont Mining Corporation. 2009. 'Community Relationships Review: Global Summary Report: March 2009.' Newmont Mining Corporation. Colorado: Newmont Mining Corporation.

Simon, M. 2009. 'The Complete History of Social Networking: CBBS to Twitter.' *MacLife*.

8 Hope for the future

The morning of Thursday 5 November 2015 was fair but hot in the Bento Rodrigues district of Mariana, southeastern Brazil. The sun shone into the valley of green and wooded hillsides. But early that same afternoon a cataclysm of ruddy mine waste surged down in a muckslide carrying mud, water and iron ore tailings in a volume estimated to be equivalent to 20,000 Olympic sized swimming pools. Within approximately half an hour, the village of Bento Rodrigues and neighbouring areas were buried. Eleven people died and, at time of writing, at least a further twelve remain missing. Hydroelectric plants were engulfed, shutting off major power sources. More than a quarter of a million people remained without potable water weeks after the event. Just over two weeks later, the waste had travelled 500 km down the Doce river, killing fish, flooding townships and emptying out into the Atlantic ocean. The environmental disaster is being called the worst in Brazil and joint mine owner BHP Billiton's history (Yolen 2015).

The responses to the Fundao and Santarem dams collapse at the Samarco Mineracao S.A. iron ore operation, jointly owned between BHP Billiton and Brazil's Vale, continued in disaster mode. Firefighters and local emergency services mounted an ongoing search for survivors. Independent assessments have been commissioned by the companies to determine reasons for the dam failure, with local and international civil society groups arguing that questions about the dam's integrity, raised only a month earlier, should have been taken more seriously. Although a 2013 report, cited by the companies, ruled out those safety issues, recent seismic activity in the area had also raised concerns and the cause of the disaster has yet to be pinpointed. Meanwhile, the Brazilian government has frozen more than \$100 million in Samarco accounts to support victim payments. BHP and Vale have signed a 'preliminary commitment' to pay out \$366 million in damages, including 'preventive emergency mitigation, repair or compensation measures'. Studies are also underway to confirm whether the sludge is toxic. More than 500 people remain homeless. The village of Bento Rodrigues is desecrated. The centuries-old Minas Gerais mining region faces a crisis of identity, as locals wrestle with the all-too-common mining dilemma of development and degradation. As Mariana mayor Duarte Junior stated in a public speech following the disaster, 'Mining represents 80 percent of our revenues. They must pay full reparations but we need the mining industry' (Yolen 2015).

From disaster, hope?

Events at Minas Gerais demonstrate that the responsible mining framework is as needed as ever. Early responses to the disaster also suggest that the timing may finally be right for such a framework's wholesale integration into the mining business. Chapter 7 theorized that a values shift or crisis-as-catalyst is required to institutionalize responsible mining. Samarco may represent both.

Responsible mining relies on several assumptions. The first is that mining, as an industry, has qualitatively changed in the twenty-first century. The theoretical work of Chapters 2 and 7 demonstrates how corporate social responsibility and its related synonyms and concerns are institutionalized within the global mining industry. The interviews and case studies throughout the book evidence an imperfect but committed attention to social performance. Returning to the Samarco dam collapse, BHP Billiton's immediate response to the Vale disaster certainly suggests a changed (or at least changing) industry. Although events continued to unfold rapidly at the time of writing, the company's earliest responses provide further illustration of an industry that is shifting attitude to social performance from one that is inherently litigious and defence-oriented towards one of acknowledging mistakes and assuming responsibilities. Consider then-BHP Billiton Chairman Brian Loton's response to Papua New Guinean litigants questioning him about the Ok Tedi case during the 1995 Melbourne annual general meeting. *The Australian* Journalist Matthew Stevens (1995) described his reply as follows:

> Loton, with typical poised precision, said BHP had satisfied its own environmental charter at Ok Tedi, promised to further improve that performance but reiterated BHP's conviction that a tailings dam could not be built at Ok Tedi. Loton also emphasized that BHP would not sell or close Ok Tedi. Loton, the silver-haired former chief executive of BHP, is one of the unchallenged giants of corporate Australia, the Man of Steel.

Compare the above to the widespread media reports of current BHP CEO Andrew Mackenzie holding back tears as he addressed the Perth annual general meeting following his trip to the Vale disaster. Far from rigid defensiveness, Mackenzie's earliest public reactions to the tragedy were requests to his advisers to get him to Brazil as quickly as possible and reports of him 'going off [the corporate] script' at press conferences (Saunders et al. 2015). Just over a decade after his Ok Tedi article and now writing for *The Australian Financial Review*, the self-same journalist Matthew Stevens headlined BHP Billiton's initial reaction, 'Mackenzie's response shines in week of gloom'. He went on to praise the CEO's gut decision to own the tragedy and head to Brazil. 'Whatever the social, commercial, diary or legal constraints might be, he and [iron ore chief Jimmy] Wilson would be in Brazil by the start of the week and he would face the media to express solidarity with those affected by the dam failure' (Stevens 2015).

Perhaps these examples are pure personality differences, and they are certainly anecdotal. But for a major miner to subjugate the traditional corporate message, even for a moment, signals an operating environment very different to the one of the late twentieth century. And this is the second major assumption on which hope for responsible mining is based. The corporate operating environment and even our understanding of what it means to be a corporation have changed. Today, activist shareholders, civil society groups, community watchdogs and consumers have more power than ever before. Tools of social media and the capability of the Internet to allow aligned but distal communities to come together to advance their causes are placing greater power in the hands of everyday stakeholders. And this is a trend that will only grow as technologies advance and users become more savvy.

Finally, and returning to the assumptions discussed in Chapter 1, the responsible mining framework assumes that we will continue to rely upon mining to support and advance the lifestyles most of us desire. But the mining industry remains exceptionally risky business. No site is failsafe, even with advances in mining technology, social performance, transformations in safety practices and prowess in construction, plant and production. Responsible mining acknowledges the choices and trade-offs necessary. It equally holds that holistic assessment, community-based agreement-making, appropriate boundaries and good governance – all shaped through processes of ethical decision-making – can reduce the social and environmental compromises necessary.

Hard questions remain

There is an optimism inherent to any discussion that dares to combine the terms 'responsible' and 'mining'. Yet current market and social environments make long-term optimism somewhat oppressive. Slogging my way through Piketty's *Capital* recently, I was staggered not only by the figures but by the evidence base behind his call for attention to the increasing disparity between the global rich and poor. These may seem abstract propositions, intellectually distanced from the pragmatics of the mine pit or processing plant, but these macro socio-economic pressures will generate real consequences for the mining and extractives industries.

So, why do we continue to question the business case for corporate social responsibility?

Chapter 2, especially, engaged this debate and I will not rehash it here. But even in conclusion, it remains a pestering question and one which I imagine will persist.

Equally persistent is the question of: why now, despite hard learned lessons from decades past, do we see companies making strategic organizational decisions that draw back on community relations resources or restructure CSR roles back into marketing and communications? At the time of writing – and as noted throughout the text – the global mining industry is entering a trough period. Stories from the field suggest that early cuts are being made into community relations and sustainable development roles, with certain majors withdrawing established social performance sections back into the archaic world of marketing and communications. Whether this will mark a trend to be bemoaned, or whether such actions will prove to be merely reactive outliers, remains to be seen.

As a reflexive researcher, I am also left pondering: Why is it that social scientists can see companies making decisions which we can predict – through evidence, modelling and experience – will lead to material losses and hinder efficiency, yet we continue – as a field – to founder in making our case about the importance of social performance? Perhaps it is our tendency to undertake our research in silos, lack of funding, fear of selling out or the drive to 'publish or perish'. Whatever the reason, the social scientific and impact assessment communities have a critical role to play in responsible mining and bear their own responsibilities in ensuring they are able to do so.

Finally, and as discussed in Chapter 3, there remains concern among certain social scientists studying community relations and social responsibility in the mining industry that innovation in social performance and related areas is dying out too soon. Attention is turning elsewhere before optimal gains have been achieved. Perhaps recent events will redouble focus and identify new champions to advance practice.

Summing up

In proposing and pursuing a framework for responsible mining, the book aims to improve social performance too frequently characterized by strong, idealized policies which are decoupled from practice; disconnections between policy requirements and on-ground resources devoted to achieving those standards; a preference towards voluntary regulatory initiatives which are relatively unenforceable but nevertheless effectively push-back against increased involuntary regulation; compliance benchmarks which appear to offer a lower threshold than that being demanded by communities at operation sites; and a privileging of legitimacy gains achieved through ad hoc programming, in lieu of more strategic activities to support longer-term community outcomes. The responsible mining framework sets out a principled and pragmatic means for addressing these concerns.

The preceding chapters have distilled global lessons learned from at least two decades of industry investment in and attention to responsible practice. It summarizes over a decade of primary and secondary social scientific research into the industry. The five principles and practices proposed in this book can foster greater industry integrity and facilitate long-term sector success.

Holistic assessment

The task of improving impact assessment's application, outcomes and usefulness is critical. From environment to health to social concerns, human rights to cultural heritage, these assessments are a major source of data for the resources industry. Holistic assessment adopts a systems lens to better capture and understand the ecosystem of impacts affecting any one project or community. It draws upon team-based transdisciplinary impact assessments, undertaken in parallel, and simultaneous analysis to generate combined recommendations that account for a comprehensive view of a mining operation's impacts and benefits.

Community-based agreement-making

Community-based agreement-making actively employs the data gathering and evidence-base of holistic assessment to ensure community ownership of and equal engagement in negotiations. It aims to balance project costs with desired community benefits. It incorporates local knowledge and concerns, while also defining the local community's economic participation in development. This secures the meaningful, continuing involvement of communities in determining their futures, relative to the development process, which includes regular monitoring, feedback and response to ongoing impacts.

Appropriate boundaries

There are considerable risks and negative consequences in the irresponsible slippage between corporate and government roles and responsibilities. Appropriate boundaries acknowledges these risks and seeks to use the strong holistic assessment evidence-base and the more symmetrical relationships produced through community-based agreement-making to reduce paternalism and dependency. Mining companies must carefully define their roles, relative to governments and communities, and align their activities with core competencies, achieving credibility through transparency.

Good governance

Any experienced resources developer is familiar with the sticky complexities of red and green tape governing the industry. Where bureaucracy or regulation is excessive, business bears the cost. Meanwhile government shoulders the burden of ensuring resources and environments are properly managed and protected. The addition of regulatory layers often creates tensions and contradictions between jurisdictions, different resource types, new technological developments and which stakeholders, concerns and agendas are prioritized.

Good governance incorporates an appropriate regulatory mix of soft and hard regulation. It accommodates transnational resource governance initiatives while encouraging legislation where necessary, especially where the rule of law is currently lacking. Good governance understands that, to paraphrase EITI founder Peter Eigen, just because an initiative is voluntary does not mean it is not mandatory once you've signed up. And it seeks the golden balance of efficiency and effectiveness.

Ethical decision-making

Finally, ethical decision-making flows throughout the responsible mining framework. It guides choices about data disclosure and citizen involvement in holistic assessment. It creates productive and respectful working relationships in community agreement-making. It assists companies to take a hard look at why they choose to provide particular infrastructure and services within operation locales and to consider whether they should. And it is intrinsic to good governance wherein compliance is a threshold but responsibility marks a gold standard.

What the responsible mining framework is…

The ideas distilled through the responsible mining framework are not revolutionary. But until now, the concept is one that has been outlined mostly in broad brush delineation of various but mostly aligned principles and guidance documents (Goodland 2012; Prospectors and Developers Association of Canada 2014; World Gold Council 2015). The framework presented here offers a more in-depth explication of the chosen principles, and those interested in responsible mining but wanting to consider different options for its construction are encouraged to refer to these guidelines. Much like CSR and sustainable development before it, responsible mining offers an important advance in approaches to principles and practice. But it is also one which is neither wholly agreed nor intractable, and multiple variations and iterations can co-exist.

The coining of 'responsible mining' is relatively new, however, with early mentions appearing around the turn of the century. Since at least 2006, the Initiative for Responsible Mining Assurance (IRMA), a multi-stakeholder network of civil society organizations, has been promoting the idea. Their draft standards were signed by Anglo-American in the United States in mid-2014. The World Gold Council and ICMM acknowledge the concept by name, and a number of mining companies and civil society groups are beginning to employ the term. It is also one that is beginning to appear in legislation, with the Canadian Parliament introducing an ultimately unsuccessful 'Responsible mining' bill in 2009 (Simons 2014). Certain academic studies are also beginning to delve into the concept, especially in relation to guiding activities in developing countries (Brown et al. 2011; Cane et al. 2015) and relative to similarly named certification schemes (Junior et al. 2015).

Initiatives including the ICMM, UNGC, the Ruggie Guiding Principles, the GRI's Mining and Metals sector supplement and the EITI set out broad values and activities that align with and buoy responsible mining. Few would deny that the ideas raised in the preceding chapters are important. Many would agree they are necessary to contemporary mining practice. The responsible mining framework, then, is an in-depth platform to launch further discussion and future practice. It is intended to raise debate and disagreement, just as it aims to draw consensus and demonstrate the necessity to integrate certain social performance concerns which may be parcelled out within practice. In so doing, it strengthens each of the framework's individual components, suggesting how the framework's employment can inform the institutional change necessary for responsible mining in the twenty-first century.

...And what it is not

Each year academics churn out an estimated 1.4 million scientific journal articles (Björk et al. 2009). Of these, only a small fraction is cited by peer researchers. Even fewer leap the chasm from scholarly output to real-world influence. This book is not an attempt to detail specific research findings or build obtuse theory. That aspect of scholarly work is valid and crucial, but it is best saved for journal articles and niche publications. The chance taken here is to introduce an overarching framework for responsible mining in the twenty-first century – introducing research-based concepts and practices to construct a scaffold onto which better, stronger, more long-term practice can be based. And I hope to have done this in an intelligently engaging manner. If this has been achieved, then the book has met its mark.

References

Björk, B.-C., A. Roos and M. Lauri. 2009. 'Scientific Journal Publishing: Yearly Volume and Open Access.' *Information Research* 14(1).

Brown, W., D. M. Franks and G. Kendall. 2011. 'The Foundation for Responsible Mining in Cambodia – Suggested Approaches.' Phnom Penh: United Nations Development Programme, 16.

Cane, I., A. Schleger, S. Ali, D. Kemp, N. McIntyre, P. McKenna, A. Lechner, B. Dalaibuyan, K. Lahiri-Dutt and N. Bulovic. 2015. 'Responsible Mining in Mongolia: Enhancing Positive Engagement.'

Goodland, R. 2012. 'Responsible Mining the Key to Profitable Resource Development: Defining "Best Practice Responsible Mining".' Institute for Environmental Diplomacy and Security at the University of Vermont, 32.

Junior, R., D. M. Franks and S. Ali. 2015. 'Designing Sustainability Certification for Greater Impact.' 54.

Prospectors and Developers Association of Canada. 2014. 'E3 Plus: A Framework for Responsible Exploration: Principles and Guidance Notes.' Toronto: PDAC, 89.

Saunders, A., J. Thomson and M. Stevens. 2015. 'How Did It Get to This?' *The Australian Financial Review*.

Simons, P. 2014. 'The Governance Gap: Domestic Laws and Other Governance Mechanisms.' *The Governance Gap: Extractive Industries, Human Rights and the Home State Advantage*. P. Simons and A. Macklin, eds. London: Routledge, 178–259.

Stevens, M. 1995. 'BHP the Tarnished Australian.' *The Australian*.

——. 2015. 'Mackenzie's Response Shines in Week of Gloom.' *The Australian Financial Review*.

World Gold Council. 2015. 'Responsible Mining.' www.gold.org/gold-mining/responsible-mining (accessed 3 March 2016).

Yolen, S. 2015. 'Fine BHP Billiton for "Scene from Hell" but Don't Shut Samarco Plead Residents.' *The Australian Financial Review*.

Appendix

Notes on primary data sources, research ethics and methods

Primary data sources

The primary research data and analyses presented throughout this book are drawn from a number of research projects completed between 2004 and 2015 in various roles at Oxfam Australia, private consulting and at The University of Melbourne. All research was completed in line with the applicable ethical codes and specific codes and requirements are noted throughout the text, as relevant. While the approach taken here is somewhat unusual, the aim of this book is to develop a conceptual framework for responsible mining, steeped in the research literature and demonstrated through a decade of research-based involvement in the industry. Risking an unconventional approach seemed worthwhile, given the singular opportunity to distil a framework from the aggregate learnings of a number of research projects.

The following pages provide a brief overview of the primary research methods used and analysis undertaken. Readers interested in deeper methodological detail may wish to refer to my own studies, cited in the reference list, or to contact me directly via Twitter: @Sara_Bice.

Beyond the business case: A new institutionalist analysis of corporate social responsibility in Australian mining, 2008–2012, The University of Melbourne

The research design for this study was divided into three phases (see Table A.1). 'Pre-fieldwork' comprised a content analysis of five major Australian mining companies' sustainability reports published between 2004 and 2008 (and covering the years 2003 to 2007) and a document analysis of industry policy documents and CSR initiatives.[1] 'Fieldwork: phase I' focused on corporate level data, consisting of a pre-interview questionnaire concerning ten pre-fieldwork identified CSR concerns and in-depth interviews with 11 senior corporate headquarter representatives of nine major Australian mining companies (see Table A.1, 'Key data sources'). 'Fieldwork: phase II' involved two operation site level case studies comprising a total of 40 in-depth interviews, a 'community life' questionnaire completed by 49 community members, site visits and observations undertaken in Australian communities which host major mining operations.

In each phase, qualitative data was analysed via thematic coding using N*Vivo 8 software (Spencer et al. 2003) (see Table A.1, 'Type of analysis performed').[2]

Descriptive statistical analysis was performed on quantitative data from the fieldwork: phase II questionnaire. Other data analysis methods used included a policy document review in the pre-fieldwork phase through which a précis of findings was completed. A review of field notes was also undertaken in relation to observations and site visits completed in phase II of the fieldwork. The intricate and thorough findings made possible through qualitative analysis facilitated the more narrative structure adopted for Parts II and III of the study and supported a discussion which is able to distil representative quotes and observations to respond to the research questions and support the research's central arguments.

The phased research design and iterative method development allowed each preceding research stage to inform subsequent stages, building a stronger method and body of data (Janesick 2003). For example, the pre-fieldwork policy document review contributed not only to identification of industry, NGO, intergovernmental agency and government priorities relative to CSR and Australian mining, but also allowed for a cross-validation of issues identified in the content analysis for purposes of building a more robust coding frame. The review also contributed to identification of ten key areas of CSR concern for Australian mining companies which informed the pre-interview questionnaire used in fieldwork: phase I. Other key outcomes from the pre-fieldwork stage included: identification of key CSR concerns of those mining companies included in the sample, a quantitative prioritisation of the major CSR disclosure topics in sustainability reports and identification of commonly deployed CSR discourses (see Table A.1, 'Key outcomes').

'Key outcomes of fieldwork: phase I' included face validation of the ten identified areas of CSR concern by experts working in the mining industry and consequent insights which informed the semi-structured interview guides used during fieldwork: phase II. Perhaps most importantly, fieldwork: phase I interviews with corporate headquarter representatives enabled examination of: common CSR discourses; processes of diffusion; central management policies and approaches to CSR; inter-organizational relationships; corporate motivational factors driving CSR; and approaches to and perceptions of voluntary and involuntary regulation.

The case studies undertaken in phase II of the fieldwork, especially semi-structured interviews, site visits and observations, secured a better understanding of: common operation site discourses and diffusion methods; on-ground management approaches; inter-company and inter-organizational relationships within communities; local motivating factors perpetuating CSR; community perceptions of company legitimacy related to CSR; and knowledge of and opinions about voluntary and involuntary regulation of companies' behaviour towards local communities and environments, including perceptions of compliance (or lack thereof). Due to a low response rate in both case study communities and the random nature of the sample, survey questionnaires did not contribute statistically significant results, but data was used to inform a broader understanding of community contexts.

Table A.1 Research design

Research phase	*Methods used*	*Level of analysis*	*Key data sources*	*Type of analysis performed*	*Key outcomes*
Pre-fieldwork	Content analysis Analysis of five major Australian mining companies' GRI Sustainability Reports	Organizational field Australian mining industry	19 reports from five major Australian mining companies covering the four years from 2003 to 2007	Thematic and textual qualitative coding using N*Vivo 8 software	• Identification of key CSR concerns of those mining companies included in the sample • Quantitative prioritization of issues reported • Identification of central CSR discourses commonly employed by the studied companies
	Document review Comprised review of core industry documents which define policies for and approaches to CSR	Organizational field Australian mining industry	Social and environmental impact studies, industry association CSR initiatives, case-based CSR literature, voluntary reporting frameworks and related international standards	Précis of major social and environmental impact studies, industry CSR frameworks and commitments, case studies, voluntary reporting frameworks and related standards	• Informed development of a priori coding frame for the content analysis • Identification of industry, NGO, intergovernmental agency and government priorities in relation to CSR and Australian mining • Cross-validation with issues identified in content analysis • Combined results of content analysis and document review resulted in identification of 10 major CSR areas of concern for Australian mining companies
Fieldwork: phase I	Pre-interview questionnaire Ranks the frequency and occurrence of 10 major CSR areas of concern for Australian mining companies, as identified through the content analysis	Formal organization Australian corporate headquarters	Questionnaire responses from 11 senior, corporate headquarter representatives of 9 major Australian mining companies	Face validation of major CSR areas of concern through interviewee feedback	• Expert confirmation that the 10 areas of concern identified in the content analysis accurately reflected corporate experience • Facilitated further shaping of the definitions of areas of concern to inform lines of questioning in Fieldwork: Phase II • Use of the instrument prior to in-depth interviews helped focus interviewees' on the research's central concerns and provided a spark for conversation, easing participants into the interviews

Research phase	*Methods used*	*Level of analysis*	*Key data sources*	*Type of analysis performed*	*Key outcomes*
Fieldwork: phase II	Community life questionnaire Quantitative survey instrument, constructed based on existing, research-validated measures of community capacity and 10 major CSR areas of concern for Australian mining companies	Informal organization Operation sites	49 questionnaire responses completed by community members at operation sites	Descriptive statistical analysis. Thematic coding of open-ended responses using N*Vivo 8 software	• Due to low response rates, no statistically reliable results were achieved • Data was used to provide a broader context to Phase II, semi-structured interview data, and to inform understanding of community contexts
	Observations/site visits	Informal organization Operation sites	Participant observation through attending various organizations' and businesses' offices, sitting in on meetings, attending community centres and events, visiting key social and recreational locations, visiting local newspaper archives and taking mine site tours where safety regulations allowed	Review of field notes	• Field notes provide important contextualization for write-up of the case studies and for better understanding participants' interview responses

Sustainability Report analysis

Content analysis encompasses an exceptionally broad range of methods, including such possible approaches as, statistical, computer-automated, network mapping, and linguistic techniques (Markoff et al. 2008). For purposes of the study, content analysis is defined as a research technique which facilitates interpretations of a prescribed data set through systematic pinpointing of particular characteristics within those selected texts (Markoff et al. 2008). Despite varied definitions, most social scientists would agree that a content analysis is primarily quantitative, methodical and systematic, and employs specifically defined and carefully followed rules, such as that all text must be coded (Markoff et al. 2008).

In brief, content analysis usually consists of the researcher choosing a specific text or set of texts and creating a 'content analysis dictionary' or 'coding frame' through which to read and analyse those texts (Stone et al. 2008). The 'language signs' named in the coding frame may represent phrases or ideas (thematic content analysis) or a particular word (textual content analysis) (Stone et al. 2008). The study examines both thematic and textual items to provide a thorough reading of major Australian mining companies' sustainability reports in Chapter 3.

The creation of the coding frame was, in itself, an important part of the content analysis. A coding frame must be constructed with consideration of the theory it aims to test (Stone et al. 2008), and should be mostly complete prior to undertaking the analysis. This a priori creation of the coding frame helps to maintain balance between objectivity and intersubjectivity (Neuendorf 2002, 11). Given that an a priori coding frame is based on knowledge deduced from prior research, the codes applied are not inordinately shaped by any one document from the data set.

The coding frame was therefore largely developed a priori, with all broad and most intermediate level categories determined prior to analysing the sustainability reports. The coding frame development was also informed by the document analysis of related publications I undertook during the research period. These documents included studies of mining's social and environmental impacts (e.g. Macdonald and Rowland 2002; Banks 2003; Vittori et al. 2006; Smith 2008), industry association initiatives (Mining Minerals and Sustainable Development 2002; International Council on Mining and Metals 2005; Minerals Council of Australia 2005; International Council on Mining and Metals 2010), case-based corporate social responsibility literature (e.g. Hilson and Murck 2000; Moon et al. 2003; Hopkins 2007; Hutchins et al. 2007) and voluntary reporting initiative frameworks and related international standards (e.g. International Organization for Standardization 2004; Global Reporting Initiative 2006a; International Organization for Standardization 2010). Sustainability reports from other industries were also examined to situate the coding frame within relevant CSR discourses (e.g. British Petroleum 2006; Johnson and Johnson 2006; Chevron 2007; Citigroup 2007; ConocoPhillips 2007).

A priori construction of the coding frame did not mean, however, that all codes were established prior to coding. An 'emergent dictionary' of select terms or themes was also established by undertaking a cursory coding of four reports from different companies from each of the years included in the sample (Neuendorf 2002, 129). I then integrated the emergent dictionary with the a priori coding frame and reapplied this finalised coding frame to all reports.

In total, over 1 million words were analysed. This included 404,957 words or approximately 1,157 pages of sustainability reporting coded using N*Vivo 8 software (2008–2012) and a further 4,319 pages were coded during the second round of the analysis (2014–2015). The content analysis explored both 'manifest' (readily observed themes consistently communicated in similar language) and 'latent' themes (Neuendorf 2002). The latent themes represented any concepts which are present but not directly measurable (Neuendorf 2002). For example, in analysing the social issues prevalent in the sustainability reports data set, report authors may not have stated explicitly that a particular case study related to issues of gender inequality or employee retention, but it was inferred that a discussion of programmes to boost women's employment or a chart containing turnover rates offered information about each of these issues, respectively. Latent themes were particularly important for the analysis, therefore, as they allowed more intangible issues to be coded individually under their objective categories and then aggregated into a somewhat more abstract, interpretative category. Definitions for both manifest and latent themes employed in the analysis are available in the coding frame, upon request.

Great effort was made to ensure that both thematic and textual code definitions had internal consistency, were mutually exclusive, and that data could be coded consistently across reports (Neuendorf 2002). While content analysis methods do rely somewhat on individual interpretations, the method's conceptual development over time has helped to reduce the degree of subjectivity of results through such practices as a priori establishment of the coding frame (Stone et al. 2008), discussed above, intercoder reliability checks or sufficient coder training (Neuendorf 2002), and the choice of an appropriately sized unit of analysis (Krippendorff 2004). Internal consistency and mutual exclusivity of code definitions and ability to code consistently across reports was ensured in the analysis.

The full coding frame is available in: Bice, S. 2011. 'On the radar? Gendered considerations in Australian mining companies' sustainability reports, 2004–2007', Lihiri-Dutt, K. (ed.) *Gendering the field: Towards sustainable livelihoods for mining communities.* Canberra: ANU E-Press or by contacting me directly via Twitter: @Sara_Bice.

Responsible Mining, 2014–2015

Research completed specifically for the production of this book was supported by a University of Melbourne, Faculty of Arts, Early Career Researcher Grant. The funding facilitated an update to the content and discourse analysis of mining company sustainability reports, attendance at major industry events, including the 6th annual EITI Global Conference, and travel for fieldwork to interview mining industry experts from around the globe. The in-depth interviews completed for this book used a semi-structured interview schedule and lasted between 50 minutes and about 1.5 hours. All interviews were transcribed and results thematically coded using N*Vivo 10 software.

Epistemology and researcher reflexivity

My use of primarily qualitative methods, especially semi-structured interviews and case studies, is strongly informed by my work's aim of garnering a more comprehensive and in-depth understanding of CSR's practical operation and its implications for communities and corporations. The research design itself echoes the *bricolage* approach which I adopt in the development of my analytical frameworks and is reflective of the common qualitative research practice in which the researcher choreographs multiple methods to facilitate a more thorough investigation of complex social concerns (Denzin and Lincoln 2003; Janesick 2003). Semi-structured interviews and case study methods incorporating observations and site visits are also firmly aligned with my pursuit of iterative, dynamic research underpinned by an ethos of interpretive study, which Robert Stake (2003, 144) describes as 'seeking out emic meanings by the people within the (study)'. Philosophically, therefore, my work tends towards the phronetic end of the social sciences spectrum, bolstered by my central interests in unpacking social motivations through asking dynamic 'how' and 'what' questions concerning corporate-community-government interactions and their implications (Flyvbjerg 2001).

My employment of semi-structured interviews as a primary means of data gathering accommodates an openness to emergent sociological texts concerning responsible mining. The method allows participants to influence and even direct conversations in a way which facilitates their capacity to tell their own stories and articulate their own understandings of and interactions with mining companies and corporate social responsibility (Legard et al. 2003; Stake 2003). Such lines of conversation facilitate consideration of issues which alternative, more stringently constructed methods may deem tangential and, therefore, irrelevant. To achieve my research aims, it is often these seemingly parenthetical discussions which, through gentle probing and reflective interpretation (Stake 2003) reveal the most novel and interesting findings.

Researcher-participant relationships

Clearly, I approached the research from a subjective position through which my academic, professional and personal experiences not only influenced the chosen research design, but also my behaviour in the field and, consequently, my analytical interpretations. For example, my prior, professional fieldwork experiences in communities located near mining operation sites throughout Australia and internationally are always in the back of my mind while undertaking new fieldwork. Professional work I have undertaken in relation to coal mining in Australia, for example, has raised my awareness of the difficult intersections between individuals' agricultural lifestyles and connections to land. Experiences in Papua New Guinea highlight issues of power imbalance, long-term development after mine closure and the difficulties of pursuing a CSR agenda where political environments are plagued by corruption. Similarly, recent experiences in Côte d'Ivoire, West Africa, make me conscious of the burdens companies may adopt when national and local governments are in violent turmoil and lack the capacity to provide even the most basic infrastructure or services, and of the severe implications for local, indigenous people when forced resettlements occur in an environment which is highly politically sensitive and lacks provision for basic human security and individual rights.

As Stake (2003, 129) notes in relation to case study methods, they 'are largely the methods of disciplining personal and particularised experience'. On the one hand, my experiences allow me to be more generally informed about the topic than I might otherwise be. They also allow me to draw on other examples to make connections between participants' individual experiences and the broader picture of corporate social responsibility in mining, as I know it to be playing out in diverse social contexts and political environments. On the other hand, my prior experiences necessarily colour my engagements with research participants to at least some degree – even as I was conscientious and attempted to forestall bias – and it is possible that my lines of questioning, style of engagement and final assessments might differ had I come to this project a *tabula rasa* on the subject.

It is also worthwhile noting that much of my prior research has adopted feminist philosophies and approaches to research methods which encourage the researcher to maintain awareness of researcher-participant power relationships while working to establish more collaborative and egalitarian research experiences (Fine 1998). While I am of the opinion that 'pure' research objectivity cannot be achieved due to the subject-nature of researchers and participants, I certainly worked to reduce discrepancies in power and consciousness between myself and participants through adopting sensitive approaches to my fieldwork and maintaining an active awareness of these dilemmas (Olesen 2000).

The qualitative methods I employ throughout my research are well-placed to address these concerns and allow me to deploy tactics to address them, including adopting an open, conversational style with participants and using broad, open-ended questions to begin discussions and allow participants opportunities to express their personal positions, experiences, views and concerns (Denzin and Lincoln 2003; Legard et al. 2003). When conducting interviews, I also adopt a reflective listening style in which I express my initial interpretations of participants' comments to ensure I have accurately interpreted their meaning and ask follow-up questions, as necessary, to achieve clarity (Legard et al. 2003). My regular practice of voice-recording and full transcription of all interviews also allows me to return to conversations to ensure I have captured participants' statements and viewpoints within context and as per their intent. Whenever practicable, I provide research participants with drafts of my writing to confirm interpretations and ensure their views are being accurately portrayed.

Notes

1 I use the term 'policy documents' as a catch-all term, encompassing industry association policies, frameworks and initiatives, intergovernmental principles and signatory initiatives, reporting guidelines, corporate policies, strategy documents and government documents, including Parliamentary inquiries and related submissions.
2 Textual coding (coding of specific words or sets of words) using N*Vivo 8 software was also performed for purposes of the pre-fieldwork content analysis of sustainability reports.

References

Banks, G. 2003. 'Landowner Equity in Papua New Guinea's Minerals Sector: Review and Policy Issues.' *Natural Resources Forum* 27(3): 223–234.

British Petroleum. 2006. 'BP Sustainability Report 2006.' London: British Petroleum.

Chevron. 2007. '2007 Corporate Responsibility Report.' San Ramon: Chevron.

Citigroup. 2007. '2007 Citizenship Report.' New York: Citigroup.

ConocoPhillips. 2007. 'ConocoPhillips Sustainable Development Report.' The Hague.

Denzin, N. and Y. Lincoln. 2003. 'Introduction: The Discipline and Practice of Qualitative Research.' *Strategies of Qualitative Inquiry*. 2nd edition. N. K. Denzin and Y. S. Lincoln, eds. Thousand Oaks, CA: Sage Publications, 1–45.

Fine, M. 1998. 'Working the Hyphens: Reinventing Self and Other in Qualitative Research.' *The Landscape of Qualitative Research*. N. Denzin and Y. S. Lincoln, eds. Thousand Oaks, CA: Sage, 130–155.

Flyvbjerg, B. 2001. *Making Social Science Matter: Why Social Inquiry Fails and How It Can Succeed Again*. Cambridge: Cambridge University Press.

Global Reporting Initiative. 2006a. 'GRI: Sustainability Reporting Guidelines.' Amsterdam: Global Reporting Initiative, 45.

Hilson, G. and B. Murck. 2000. 'Sustainable Development in the Mining Industry: Clarifying the Corporate Perspective.' *Resources Policy* 26(4): 227–238.

Hopkins, M. 2007. *Corporate Social Responsibility and International Development: Is Business the Solution?* London: EarthScan.

Hutchins, M. J., C. L. Walck, D. P. Sterk and G. A. Campbell. 2007. 'Corporate Social Responsibility: A Unifying Discourse for the Mining Industry?' *Greener Management International* 52: 17–30.

International Council on Mining and Metals. 2005. 'Reporting Against the ICMM Sustainable Development Principles: Resource Guide to Assist ICMM Members Meet Their Reporting Commitments.' London: ICMM, 20.

——. 2010. 'Sustainable Development Framework.' www.icmm.com/our-work/sustainable-development-framework (accessed 1 April 2016).

International Organization for Standardization. 2004. 'ISO 14001:2004 Standards Catalogue: Mining and Minerals.' Geneva: ISO.

——. 2010. 'ISO 26000 Social Responsibility Standard.' Geneva: ISO.

Janesick, V. J. 2003. 'The Choreography of Qualitative Research Design: Minuets, Improvisations and Crystallization.' *Strategies of Qualitative Inquiry*. 2nd edition. N. K. Denzin and Y. S. Lincoln, eds. Thousand Oaks, CA: Sage Publications, 46–79.

Johnson and Johnson. 2006. 'Sustainability Report.' New Brunswick: Johnson and Johnson, 32.

Krippendorff, K. 2004. *Content Analysis: An Introduction to Its Methodology*. 2nd edition. Thousand Oaks, CA: Sage.

Legard, R., J. Keegan and K. Ward. 2003. 'In-Depth Interviews.' *Qualitative Research Strategies: A Guide for Social Science Students and Researchers*. 3rd edition. J. Ritchie and J. Lewis, eds. London: Sage Publications Ltd, 138–169.

Macdonald, I. and C. Rowland. 2002. 'Tunnel Vision: Women Mining and Communities.' Fitzroy: Oxfam Community Aid Abroad, 51.

Markoff, J., G. Shapiro and S. Weitman. 2008. 'Toward the Integration of Content Analysis and General Methodology.' *Content Analysis*. R. Franzosi, ed. Los Angeles: Sage, 268–311.

Minerals Council of Australia. 2005. 'Enduring Value: The Australian Minerals Industry Framework for Sustainable Development.' Kingston: Minerals Council of Australia, 20.

Mining Minerals and Sustainable Development. 2002. 'Breaking New Ground: The Report of the MMSD'. Earthscan.

Moon, J., A. Crane and D. Matten. 2003. 'Can Corporations Be Citizens? Corporate Citizenship as a Metaphor for Business Participation in Society.' *International Centre for Corporate Social Responsibility Research Paper Series* 32.

Neuendorf, K. 2002. *Content Analysis Guidebook*. Thousand Oaks, CA: Sage.

Olesen, V. 2000. 'Feminisms and Qualitative Research into the New Millennium.' *The Handbook of Qualitative Research*. N. Denzin and Y. S. Lincoln, eds. London: Sage, 215–255.

Smith, G. A. 2008. 'An Introduction to Corporate Social Responsibility in the Extractive Industries.' *Yale Human Rights and Development Law Journal* 11.

Spencer, L., J. Ritchie and W. O'Connor. 2003. 'Analysis: Practices, Principles and Processes.' *Qualitative Research Strategies: A Guide for Social Science Students and Researchers*. 3rd edition. J. Ritchie and J. Lewis, eds. London: Sage Publications Ltd, 199–218.

Stake, R. 2003. 'Case Studies.' *Strategies of Qualitative* Inquiry. 2nd edition. N. K. Denzin and Y. S. Lincoln, eds. Thousand Oaks, CA: Sage Publications, 134–164.

Stone, P., D. Dunphy, M. Smith and D. Ogilvie. 2008. 'The Construction of Categories for Content Analysis Dictionaries.' *Content Analysis*. Roberto Franzosi, ed. Los Angeles: Sage, 128–155.

Vittori, L., S. Martin and S. Bice. 2006. 'Mining Ombudsman: Case Updates 2005.' *Mining Ombudsman Case Reports*. Melbourne: Oxfam Australia. 20.

Index